# How Not to Be Eaten

# How Not to Be Eaten

*The Insects Fight Back*

Gilbert Waldbauer

*With illustrations by James Nardi*

UNIVERSITY OF CALIFORNIA PRESS

*Berkeley Los Angeles London*

University of California Press, one of the most
distinguished university presses in the United States,
enriches lives around the world by advancing scholarship
in the humanities, social sciences, and natural sciences.
Its activities are supported by the UC Press Foundation
and by philanthropic contributions from individuals and
institutions. For more information, visit www.ucpress.edu.

University of California Press
Berkeley and Los Angeles, California

University of California Press, Ltd.
London, England

Library of Congress Cataloging-in-Publication Data

Waldbauer, Gilbert.
    How not to be eaten : the insects fight back / Gilbert
Waldbauer ; with illustrations by James Nardi.
        p.   cm.
    Includes bibliographical references and index.
    ISBN 978–0–520–26912–5 (cloth : alk. paper)
    1. Insects—Defenses.   2. Insects—Predators of.
I. Title.
    QL496.W336   2012
    595.7—dc23

                                        2011024488


Manufactured in the United States of America

21   20   19   18   17   16   15   14   13   12
10   9   8   7   6   5   4   3   2   1

In keeping with a commitment to support
environmentally responsible and sustainable printing
practices, UC Press has printed this book on Rolland
Enviro100, a 100% post-consumer fiber paper that
is FSC certified, deinked, processed chlorine-free,
and manufactured with renewable biogas energy.
It is acid-free and EcoLogo certified.

*To Nancy Clemente*

*Dear friend and the best and most helpful*
*editor I have known*

# CONTENTS

# PROLOGUE

All animals must eat. If they don't, they cannot fulfill the three basic imperatives of life: to *grow* and to *survive* long enough to *reproduce*. But who eats whom, and why? Except for climatic factors such as droughts or freezing temperatures, predators are probably the most pervasive and dangerous threat to the survival of most animals. Because all organisms require food, the relationships between the eaten and the eaters are a—perhaps *the*—central aspect of what goes on in a community of organisms, which, together with their physical environment, constitute an ecosystem.

In almost all land and freshwater ecosystems, insects are the most abundant animal food. Without a doubt, knowing the ways in which insects avoid becoming a meal for an insect-eating predator and the ways in which predators evade their defensive strategies is essential to understanding how ecosystems work. Moreover, these relationships are in themselves fascinating, sometimes bizarre, and always enlightening.

Many, many different kinds of organisms make a living by preying on insects. They include a few plants, but most of them are representatives of virtually all the major animal groups (classes)—except those that live only or mainly in the seas: sponges, jellyfish, starfish, clams, and snails—ranging from spiders and insects to vertebrates such as lizards, birds, and mammals.

Natural selection favors changes (mutations) that in one way or another improve an organism's ability to cope with its environment, to better exploit opportunities and avoid being eaten by a predator. "During evolution," as J.R. Krebs and N.B. Davies wrote, "we expect natural selection to increase the efficiency with which predators detect and capture prey. On the other hand, we would also expect selection to improve the prey's ability to avoid detection and to escape. The complex adaptations and counter-adaptations we see between predators and their prey are testament to their long coexistence and reflect the result of an arms race over evolutionary time."

These adaptations and counteradaptations are multitudinous, diverse, and sometimes so extraordinary that they defy belief. A spider lures certain male moths to their deaths by counterfeiting the chemical sex attractant of females of the victims' species. A burrowing owl uses bait to attract the beetles that it eats. A praying mantis attracts its prey, nectar-seeking bees and flies, by masquerading as a large, colorful flower. Some ant lions and a few other insects dig pitfall traps and quietly wait, unseen, in the bottom to devour careless insects that stumble into the pit.

A few caterpillars hoodwink insectivorous birds by posing as repulsive bird droppings. Certain nerve fibers of some fleet-footed insects are greatly enlarged to hasten the arrival at the central nervous system of nerve impulses, generated by the perception of a warning signal, that will trigger flight. Most moths are active only at night, and during the day many hide in plain sight, not making the slightest movement as they cling to the bark of a tree or some other surface that matches their remarkably deceptive camouflage. Equally astonishing are the strategies of some predators. For instance, as Malcolm Edmunds wrote, the Old World birds known as bee-eaters "defang" a stinging insect by rubbing its abdomen against its perch to "squeeze the venom out of its sting gland."

The ten chapters that follow elucidate the strategies and counter-strategies in the everlasting evolutionary arms race between predators and prey.

### NOTE TO READER

As you read this book, you will come across quotations from or references to the work of scientists and other writers. It is only fair that you have the opportunity to read and evaluate these publications on your own. In Selected References at the back of the book, you will find bibliographic citations, listed by author, that will lead you to these published sources for each chapter. People who provided me with unpublished information are identified in the text but are not listed in Selected References.

# ACKNOWLEDGMENTS

I am greatly indebted to the many friends and colleagues whose support and expertise improved this book, making it much better than it would have been without their help. Outstanding among them are my wife, Phyllis Cooper Waldbauer, and my dear friend Nancy Clemente. Phyllis read several drafts of the manuscript and made many constructive suggestions. Nancy, who masterfully edited the five of my books published by Harvard University Press and who is now retired, volunteered to edit this manuscript pro bono. Many others freely shared their scientific expertise: May Berenbaum, the late John Bouseman, Lincoln Brower, Sydney Cameron, Larry Hanks, Steven Malcolm, James Nardi, David Seigler, and James Sternburg. With great patience, Karen Trame and Elizabeth Berry typed the manuscript. My agent, Edward Knappman of New England Publishing Associates, was, as always, very helpful.

# ONE

# Insects in the Web of Life

Insects constitute by far the largest amount of animal food available to flesh eaters both on dry land and in freshwater. The one quarter of the earth that is not covered by the oceans and seas is inhabited by an immense and not yet completely censused population of insects. The 900,000 currently known insect species (at least three million are yet to be discovered and named, according to reasonable estimates [Stephen Marshall]) constitute about 75 percent of the currently known 1,200,000 animal species on land, in freshwater, and in the oceans. The Canadian entomologist Brian Hocking made the daring but educated guess that the world population of insects is about one quintillion (*1* followed by eighteen zeroes) individuals. Even if he overestimated by trillions, that would still be a stupendous population.

Although insects are small, they are generally so numerous in most terrestrial and freshwater ecosystems that, on a per-acre basis, they not only outnumber but also outweigh all the other animals—including deer and moose—combined. On the face of it, this is hard to believe. But keep in mind that a single acre of land may be home to many millions of insects of hundreds or even thousands of species. By contrast, the home territory of one small bird is likely to encompass as much as an acre, and that of a large mammal, such as a thousand-pound moose,

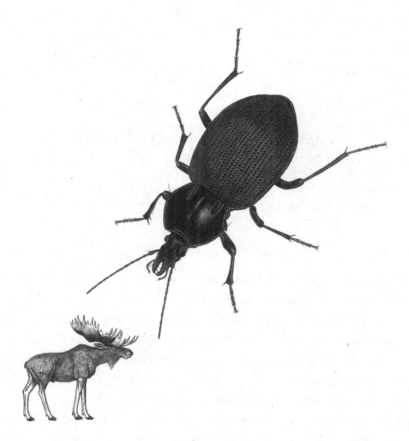

Figure 1. A thousand-pound moose (representing all mammals, birds, and other vertebrates in its ecosystem) and a tiny beetle (representing all insects in the same ecosystem) drawn to a scale representing their biomass as pounds per acre.

several hundred or even thousand acres. Thus the biomass of an animal that weighs hundreds of pounds may be much less than one pound per acre. Also keep in mind that most people notice only a few of the many insects around them, perhaps a ladybird beetle or a large and beautiful butterfly, but more often the insects that sting, bite, or otherwise annoy them. Yet the other insects, by far the vast majority in almost any ecosystem, go unnoticed. Not only are they small, but many are difficult to see because they are camouflaged, and many are out of sight because they live in the roots, stems, or other parts of plants; as parasites within the bodies of insects and many other animals; or in the soil or other cracks and crevices of the environment.

Insects are, either directly or indirectly, the most plentiful source of flesh for animals that don't eat plants. But they are important to these predators not just because of their abundance. Plant-feeding insects, estimated to be about 450,000 species, and the insects and other animals that eat them are by far the most important link between green plants and animals that don't eat plants, a conduit through which predators receive the energy of the sun, which green plants—and only green plants—can capture and make available to animals via photosynthesis, in the form of sugars. Insect-eating insects play another significant, although less important, role. By eating tiny organisms and incorporating their prey's nutrients in their own bodies, large insects become "nutrient packages" for large insectivores that cannot profitably pursue and eat tiny organisms themselves.

Data gathered by Eugene Odum and other ecologists show just how important a part of the food chain insects are in specific ecosystems. For example, in a field of herbaceous plants in North Carolina, the biomass of the plant-feeding insects alone—not including any predaceous, parasitic, or scavenging species—was nine times greater than that of sparrows and mice, the larger and more conspicuous and by far the most numerous of the vertebrates in that field. On an East African plain, just two species of ants—only those two, among hundreds of other kinds of insects—were about equal in weight, per acre, to the combined weight of the large grazing animals, such as wildebeests, zebras, and antelopes. In these two habitats and in almost all others, insects are by far the most abundant of the prey animals in both numbers and biomass. As is to be expected, and as we will see in the next chapter, hundreds of thousands of different kinds of animals exploit this nutritious, protein-rich food: spiders, scorpions, insects, frogs, toads, lizards, birds, mammals.

The insects almost certainly have more different lifestyles, ways of surviving and "making a living," than do any other group of animals. One species or another occupies every—or nearly every—ecological niche. An ecological niche is not just a place; it includes all of the resources, food, nesting sites, hiding places, and so on, required by an organism. Except for aquatic species, insects that undergo gradual metamorphosis occupy essentially the same niche throughout their lives. Those with complete metamorphosis often occupy two very different niches in their larval and adult stages.

Dragonflies, grasshoppers, cockroaches, mantises, true bugs (order Hemiptera), and lice are some insects that gradually metamorphose. A newly hatched grasshopper—a nymph—looks very much like its parents but lacks wings. As it grows, it molts several times, and its developing wings, which are external, can be seen gradually increasing in size until the hopper stops growing and molts for the last time to become an adult with flightworthy wings. Insects with gradual metamorphosis have only three life stages: the egg; the nymph, the growing stage; and the adult, the egg-laying reproductive stage. Nymphs look and behave much like adults, except in most aquatic species. For instance, adult dragonflies are aerial acrobats that pursue flying insects. But the nymphs are aquatic, don't look at all like the adults, and are fierce predators that eat aquatic insects and even small fish.

Beetles, fleas, flies, wasps, bees, moths, and butterflies are among the many insects that undergo a complete metamorphosis. The baby butterfly just hatched from the egg is a wormlike larva that does not at all resemble its parents. A biologist from another planet might think that the larva and the butterfly are two quite different kinds of animals, as dissimilar as birds and snakes. Complete metamorphosis proceeds in four life stages: the egg; the wingless larva, called a caterpillar in the case of butterflies and moths; the pupa, the transition stage in which the larva metamorphoses into the adult; and the winged reproductive stage. Larvae not only look different than their parents but also usually behave very differently. Caterpillars, for example, have chewing mouthparts, and most feed on plants, usually the leaves. The wingless pupae, with only a few exceptions, can squirm but cannot walk or crawl and are usually tucked away in a safe place, perhaps in the soil, under bark, or in a silken cocoon. The adult, a butterfly or moth for example, has large wings that developed internally in the pupa, as do its long, soda straw–like siphoning mouthparts, used for sucking nectar from blossoms.

The larvae and the adults are specialists, anatomically and behaviorally equipped to do particular tasks. The larvae eat, grow, and do their best to foil predators. The adults suck the sugary nectar that supplies energy to fuel their frequent flights as they seek mates and as the females distribute their eggs. Most butterflies and moths, as well as many other insects, glue their eggs only to one of the few plant species their fussy—host plant–specific—offspring will be willing to eat. The larvae are "eating machines," and the adults are flying gonads or, as Carroll Williams said, "flying machines devoted to sex." Because of the benefits of such specialization, complete metamorphosis has been an evolutionarily more successful strategy for survival than gradual metamorphosis, at least as judged by the number of known insect species on earth today—only about 135,000 (15 percent) with gradual metamorphosis, while 765,000 (85 percent) have complete metamorphosis.

Insects with either type of metamorphosis can occupy similar ecological niches. Grasshoppers, Japanese beetles, June beetles, and many other insects occupy fairly commonplace niches. The adults feed on foliage and lay their eggs in the soil. However, after hatching from the egg, nymphal grasshoppers immediately make their way to the surface and feed on the leaves of grasses or herbaceous plants, while the white C-shaped larvae, known as grubs, of the two beetle types, which both have complete metamorphosis, remain in the soil and feed on plant roots. They pupate in the soil, and the adults dig their way up to the surface after they have shed the pupal skin.

Other insects have more complex and elaborate lifestyles. For example, a male burying beetle whose search for a dead animal has been successful sits on the body of the small dead animal he has discovered and releases a scent, a sex-attractant pheromone. A female soon joins him, and, working together, they burrow back and forth beneath the dead body until it sinks deep enough into the ground so that they can cover it with soil. Then they create an open space underground around the buried carcass and cover the dead animal with a secretion that kills bacteria, thereby delaying decomposition. The female then lays as many as thirty eggs in the soil near the carrion. After the larvae hatch, they crawl to a nest prepared by their parents, who feed them by regurgitating predigested carrion. Eventually the larvae feed on their own. The father then leaves, but the mother guards her young until they are ready to pupate. In a few weeks her offspring emerge from the soil as adults and begin another cycle.

Elsewhere, a tiny female gall wasp inserts an egg into an oak leaf. As May Berenbaum wrote in *Bugs in the System,* gall makers commandeer "the plant's hormonal system in such a way that the plant is induced to produce bizarre and unusual growths [galls], which provide the insect with a place to live and with nice nutritious tissue on which to feed." The larva that hatches from her egg causes the oak leaf to form an abnormal growth, in this case an "oak apple," a light tan spherical gall that may be as large as a table tennis ball. Another gall maker, a moth, lays an egg in the stem of a growing goldenrod in spring, causing the plant to produce an egg-shaped thickening almost an inch in diameter. In summer the caterpillar grows to full size. The following spring it gnaws an exit hole through which, after it pupates, it will emerge as a moth. But it may not survive that long. In winter a hungry downy woodpecker may peck a hole in the gall, pull out the caterpillar, and make a meal of it.

These few examples give no more than an inkling of the many different ways in which insects conduct their lives. Insectivores must, of course, have the anatomical and behavioral adaptations required to catch their prey. A bird, for example, can snatch an adult grasshopper, beetle, gall wasp, or gallfly from the air with its beak, but only a

tunneling animal or a bird that probes in the soil is likely to find sub-
terranean eggs, grubs, or pupae. Only a woodpecker is likely to get at a
larva in a gall or burrowing under the bark of a tree.

The evolution of the millions of different kinds of insects that live on
earth now and the many extinct species that we know only as ancient
fossils began about four hundred million years ago, when the first
insects-to-be were gradually leaving the water, where life began, to
move onto the land. They probably reached the shore via moist organic
debris at the edges of freshwater ponds and once on land probably
continued to feed on soft rotting organic matter, which they ate with
their primitive, unspecialized mouthparts, the organs of ingestion.
From these simple creatures evolved the diverse assortment of modern
insects, as different from one another as grasshoppers with mouthparts
specialized for chewing on plants, butterflies with tubelike mouthparts
for sucking nectar from flowers, and mosquitoes with piercing-sucking
mouthparts for consuming the blood of birds, mammals, or reptiles.

Plants and animals, of course, continue to evolve. But how does
evolution work? Charles Darwin had the brilliant insight that natural selection is the driving force of evolution, producing new species
just as breeders produce new dog breeds through artificial selection,
by selecting animals with desirable traits to be the parents of the next
generation. (Keep in mind that all breeds, from the tiny Chihuahuas to
the huge Saint Bernards, are descended from the wolf.) Natural selection, while tending to cull poorly adapted individuals, favors those
better adapted to avoid hazards and to take advantage of opportunities. For example an individual with even slightly better camouflage
than others will be somewhat less likely to be noticed by a predator
and, consequently, somewhat more likely to survive and become a parent. Heritable adaptive traits are passed on to future generations and
given enough time will spread to all members of a population. As the
centuries or millennia pass, more favorable mutations accumulate in a
population until those who have them are so different from the other

members of their species that they become a separate, distinct, reproductively isolated species, one whose members do not breed with members of other species.

These new adaptive traits constantly arise as genetic mutations caused by means such as radioactivity, ultraviolet light, cosmic rays, or intrinsic factors in DNA, the genetic material itself. Mutations are random, some favorable and many unfavorable. However, evolution is by no means a random process; it is directed by natural selection, which tends to eliminate unfavorable mutations and generally perpetuates favorable mutations. Think of a prospector panning for gold. He scoops up a mixed assortment of sand, pebbles, and—with luck—a few bits of gold. But only the heavier flakes and nuggets of the valuable gold survive the panning. They are not, unlike the lighter, valueless mixture of sand and gravel, washed out of the pan as he swirls the water. In a similar way, natural selection preserves favorable genes and eliminates deleterious genes.

Some of the insects' most important adaptations are responses to insectivores, a numerous and pervasive threat to their survival. The ultimate goal of any organism is, of course, to reproduce itself, to pass its genes on to future generations, and to accomplish this it must survive long enough to attain sexual maturity. As the great English naturalist Henry Bates wrote in 1862 in "Contributions to an insect fauna of the Amazon Valley, Lepidoptera: Heliconidae":

Every species in nature may be looked upon as maintaining its existence by virtue of some endowment enabling it to withstand the host of adverse circumstances by which it is surrounded. The means are of endless diversity. Some are provided with special organs of offence, others have passive means of holding their own in the battle of life. Great fecundity is generally of much avail. . . . A great number have means of concealment from their enemies, of one sort or another. Many are enabled to escape extermination or obtain subsistence, by disguises of various kinds: amongst these must be reckoned the adaptive resemblance of an otherwise defenceless species to one whose flourishing race shows that it enjoys peculiar advantages.

The last sentence refers to the fascinating subject of the last chapter of this book, harmless insects, and a few other harmless animals, that foil predators by bluffing, mimicking the appearance and even the behavior of other insects or other animals that sting, are unpalatable, or are avoided by predators for other reasons.

Besides reproducing themselves, insects perform indispensable ecological services. As discussed above, they are the most important link between plants and animals that don't eat plants, and they have other important roles in virtually all terrestrial and freshwater ecosystems. One of their major functions, which we have all heard about, is to pollinate plants. Most of the green plants are flowering plants, called angiosperms (Greek for "a seed encased by an ovary"), and except for hummingbirds, bats, and just a few other animals, it is the insects that transport the sperm-containing pollen from the male parts of one flower to the female parts of another. Most flowers have coevolved with bees, butterflies, or other pollinators. Their colors and scents attract insects and reward them with nectar and pollen, which many insects eat and which are virtually the only foods consumed by the thousands of species of bees (at least 3,500 in North America alone). No one knows how many of the flowering plant species are pollinated by insects, but Stephen Buchmann and Gary Nabhan have reported that of the 94 major crop plants on earth, the wind pollinates 18 percent, insects 80 percent, and birds 2 percent.

Insects have many other functions in the web of life, only a few of which I will mention here. Plant-feeding insects help to keep plant populations from increasing to a size that would disrupt a stable ecosystem. For example, when the European Klamath weed, also known as St. John's wort or locoweed, reached California, its population exploded because it had no natural enemies there; it choked out grasses in pastures to the extent that they were useless for grazing cattle. After a European leaf beetle that eats Klamath weed was introduced into California, the weed became scarce, and grew mainly in shady places, where it was less likely to be attacked by the leaf beetle. An agricultural entomologist remarked that insects are their own worst enemies. And indeed they are. Thousands of insects, probably more than 300,000 of the known species, eat other insects. As Peter Price noted, insects, mostly ants, are the "world's premier soil turners," more so than earthworms, which are generally given credit for this. Without the scarabs and other dung-feeding insects, we might, to use a bit of hyperbole, be knee-deep in excrement. Furthermore, ants and other insects disperse the seeds of some plants.

In the next chapter we will meet a few of the many animals—spiders, scorpions, toads, birds, bats, mice, and even bears—that eat insects. The threat to the insects from these insectivores is enormous, but as we will see in following chapters, insects have evolved many, often amazing ways to avoid being eaten. But keep in mind that neither insects nor other organisms are completely immune to predation. If they were, their populations would probably explode, causing ecological havoc.

# TWO

# The Eaters of Insects

In the middle of the night, a little bolas spider hangs from a plant by a few threads of silk. When a flying moth comes close enough, the hungry spider flicks at it a length of silk thread with a droplet of very sticky glue at the end. This is its bolas. With a bit of luck, the glue catches the moth, which the spider reels in and makes a meal of. The bolas spider's weapon is unique among animals—except for humans, who have invented similar weapons, the lasso and the Argentine gaucho's bolas, for which the spider is named. The spider doesn't just wait for an insect to wander along. It lures in its prey, which are always moths, by means of a false signal, a counterfeit version of the odorous sex-attractant pheromone released by the female moth to attract a mate. The clue that led to the discovery of this surprising chemical subterfuge is that these spiders catch only male moths of one species.

The bolas spider is just one among hundreds of thousands of animals that eat insects and is by no means the only one that has evolved an ingenious way of capturing its prey. The most abundant of the insectivores are themselves insects, at least three hundred thousand species. The other insectivores are far fewer, but they run the gamut of the animal kingdom: spiders, scorpions, centipedes, fish, frogs, toads, salamanders, turtles, young alligators, lizards, snakes, birds such as

Figure 2. A pair of burrowing owls have placed clumps of cow manure around the entrance to their burrow, a bait that attracts dung beetles, which the owls will eat.

woodpeckers, nuthatches, swifts, swallows, and warblers, and mammals such as anteaters, armadillos, skunks, shrews, mice, bats, and even bears.

The eaters of insects have evolved innumerable tactics and strategies for finding, capturing, and consuming their share of this great multitude of creatures. Dragonflies course over ponds, capturing mosquitoes in a basket formed by their spiny legs. Well-camouflaged praying mantises sit motionless as they patiently wait to grab passing insects with their raptorial front legs. Among the birds, colorful warblers flit from twig to twig, rapidly scanning leaves for caterpillars; flycatchers dart from their perches to snap up flying insects; and blue jays sometimes scan tree trunks for camouflaged moths that blend in with the bark on which they rest during the day. Bats employ sonar—echolocation—to find flying insects at night. An armadillo uses its long, sticky tongue to capture the ants, beetles, and other insects it uncovers as its rather long snout furrows through leaf litter and loose soil. We usually think of squirrels as vegetarians, but William Burt watched a thirteen-lined ground squirrel dig in the soil for white grubs (the larvae of a June beetle). The bolas spider's tactics are unique, but other spiders use a variety of quite different hunting strategies. Among them are spiders that spin the familiar webs of sticky silk that snare flying insects, wolf spiders that chase their prey, and spiders that lurk in tunnels covered by trapdoors and dash out to grab and envenom passing insects that stumble into their trip wire, a single strand of silk.

The familiar flat, circular webs spun by the orb-weaving spiders are sticky lacework nets designed to catch unwary insects that blunder into them. Although they are marvels of engineering, they are also very beautiful, poems in symmetry—especially in early morning, when they glitter with little drops of dew. Long, straight strands of silk, which are not sticky, radiate from the hub of the web like the spokes of a wheel. Interspersed among them are long strands of exceedingly sticky silk that form a sequence of closely spaced loops spiraling from the hub of the web to its outer edge.

When the spider, waiting motionless on the hub, feels the vibrations caused by a struggling insect stuck to the web, it plucks the radial threads one by one, "apparently," wrote Rainer Foelix, an expert on spider biology, "to probe the load on each radius. In other words, it tries to find the exact position of the prey." It then "will rush out of the hub using exactly that [nonsticky] radial thread which leads to the prey." Only after it has wrapped the prey in silk does it administer its venomous bite. The spider "then cuts the neatly wrapped 'package' from the web and carries it to the hub. There it is attached by a short thread before it is eaten."

Few insects other than moths, as Thomas Eisner and his coworkers discovered, manage to escape from the webs of orb-weaving spiders. Moths are sometimes saved by the tiny, easily detached scales that cover their wings and bodies. (The colored "powder" that clings to your fingers when you handle a moth or a butterfly consists of these scales.) Moths that blundered into a web, Eisner noted, "were detained only momentarily, and usually flew off seemingly unaffected by the encounter. However, they invariably left behind, stuck to the particular viscid threads . . . that bore the impact, some of the scales that ordinarily cover their wings and bodies. . . . Coated with scales, the threads are no longer adhesive, and the moth is free to escape."

On a sunny June day in northern Michigan, I spotted an interesting-looking insect on a large white blossom of a Canada anemone. (It turned out to be a fly that mimics a yellow jacket wasp. You will read a lot more about mimicry in later chapters.) This insect, oddly enough, wasn't moving at all, and its posture seemed unnatural. Looking more closely, I saw that it was clutched by a white crab spider that was all but invisible on the flower. Many crab spiders, like this one, are ambushers that lurk motionless in a blossom as they wait for their prey to land—a fly, bee, or other insect. Some can, in the course of about a week, change from white to yellow or yellow to white to match the color of the blossom on which they are lurking. "Fortified with extremely potent venom in compensation for weak Chelicerae [pincers], the small crab spiders," Foelix commented, "are formidable creatures that attack insects and other spiders much larger than themselves."

The fact that these spiders evolved the ability to change their color suggests that at least some insects are wary and will not land on a flower occupied by something that might be a predator. In *The World of Spiders,* W. S. Bristowe described a simple experiment which showed that this seems to be so. On half of sixteen yellow dandelion blossoms that he had placed on a lawn, he put a black pebble about the size of a crab spider; on the other half he placed a pebble that was about the same size but matched the dandelion blossom's color. As Bristowe watched these blossoms for half an hour, only seven insects visited the ones with black pebbles, while fifty-six flies and bees visited the blossoms with yellow pebbles.

Some insects, such as the familiar praying mantis, are predators that, like a tiger or a wolf, during their lives will capture many prey animals and more or less immediately kill and eat them. Other insects, mainly many of the wasps and all of the flies of the family Tachinidae, according to Richard Askew, are parasites that develop from egg to full-grown larva within the body of just one host, usually an immature insect—perhaps a beetle grub, a caterpillar, or a grasshopper. However, parasitic insects are best referred to as *parasitoids.* True parasites, such as the intestinal worms of humans and other mammals, usually do not kill their hosts. But although a parasitoid begins relatively benignly, just absorbing nutrients from its host's blood, it eventually becomes a predator, killing its host by devouring its tissues.

Among the many thousands of insects that parasitize other insects, the wasps of the genus *Trichogramma* (they have no common name) are particularly interesting because of their habits and because agriculturalists disperse them in fields to protect various crops from leaf-eating caterpillars. These tiny wasps—the largest are only about 0.04 inches long—insert their eggs into the eggs of many different kinds of moths and butterflies. The *Trichogramma* larvae destroy the host egg by consuming its contents. The size and even the anatomy of the adult parasite that emerges from the egg, Askew explained, varies with the size and species of the host egg. Wasps reared from the large eggs of a cutworm moth were nearly twice the size of those of the same species reared from the much smaller eggs of a grain-feeding moth. Males of another species reared from a moth egg had normal wings, but those reared from an alderfly egg had no wings at all. *Trichogramma* wasps can be raised by the millions in the eggs of a domesticated colony of grain moths. If a crop, perhaps cotton, is likely to be seriously damaged by the caterpillar progeny of egg-laying moths, *Trichogramma* are often released in the field in the form of larvae developing in grain moth eggs that had been glued to slips of paper by their mothers.

Many nonparasitic insects that feed on other insects are hunters that actively pursue their prey. Robber flies (family Asilidae) dart from a perch, perhaps the tip of a twig on a low shrub, to snatch flying insects from the air. When the fly comes back to land, it sucks its victim dry after injecting it with a secretion containing poisons that kill it and enzymes that liquefy its inner tissues. Almost all of the more than 2,600 species of ground beetles in North America are insectivores, such as the colorful, inch-long, tree-climbing caterpillar hunters (genus *Calosoma*). Adult cicada killers (*Sphecius speciosis*)—wasps barred with yellow and black and about 1.5 inches long—are vegetarians that sip nectar from flowers but search trees for the dog-day cicadas that they will feed to their larvae ensconced in underground cells.

Ambushers, by contrast, are stealthy insects that are generally well camouflaged and sit motionless as they wait patiently for their prey to come along. Like the white crab spider, ambush bugs (family Reduviidae) lurk on flowers, often goldenrods, waiting to snatch up with their raptorial front legs a visiting nectar feeder, such as a bee, a wasp, or a butterfly. As do all of the other true bugs, they have piercing-sucking mouthparts, which the nonvegetarians use to suck their prey dry, rather than masticate it like the mantises and other insects with chewing mouthparts.

One of the most deceptive of the ambushing insects is a southeast Asian praying mantis (*Hymenopus bicornis*) that masquerades as a pink flower of the straits rhododendron, a shrub known as the Sendudok in the Malay language. Hugh Cott summarized Nelson Annandale's 1900 account of the appearance and behavior of this remarkable insectivore in the report of the University of Cambridge's expedition to the Malay Peninsula. The bright pink mantis, the color of the blossoms among which it rests, is patient and motionless as it waits for its next meal to come close enough to be snatched. Its deception is enhanced by wide .

pink petal-like flanges on its middle and hind legs. Insects, as Wolfgang Wickler noted, actually land on the mantis's body and probe for nectar, "for which they pay with their lives." Its "alluring colouration," as Hugh Cott put it, is a bait that attracts the insects that it eats. This is one of many examples of aggressive mimicry, the duping of potential victims, causing them to relax their guard. The mantis's disguise may also save its life. Insectivores such as lizards and birds are likely to pass it up because it is so deceptively camouflaged. 

The mantis's ruse is most effective when it lurks among blossoms that it resembles. Consider Nelson Annandale's account of the dogged persistence of one such insect as it searched for an appropriate resting place. A captive mantis put on the ground near a large branch of a Sendudok

> deliberately walked towards the branch, swaying its whole body from side to side as it progressed, and commenced to climb one of the twigs. This twig, however, bore only green buds and unripe fruit. When the Mantis reached the tip of the twig and found no flowers, it remained still for a few seconds, and then turned and descended with the same staggering gait. It proceeded to climb another twig. This also bore no flowers. The Mantis descended from it and mounted a third twig which was topped by a large bunch of full-blown blossoms. To these it clung by means of the claws of the two posterior pairs of limbs. For a few minutes it remained perfectly still, and then began swaying its body from side to side, as it had done while walking.

Annandale, quoted by Cott, tells us how well this mantis's aggressive mimicry deceives the insects that it will eat.

> Almost as soon as the Mantis had settled itself on the inflorescence, a small, dark . . . [fly] of a kind very commonly seen on the flowers of this species of [shrub] alighted on one of its hinder legs. It was soon joined by others, apparently of the same species as itself. They settled quite indiscriminately on the petals and on the body and limbs of the Mantis. . . . The Mantis made no attempt either to drive off or to capture the small flies, for its motions seemed to attract rather than to repel them. After a short

time a larger [fly,] as big as a common house-fly, alighted on the inflorescence within reach of the predatory limbs. Then the Mantis became active immediately; the fly was seized, torn in pieces and devoured, notwithstanding the presence of a large crowd of natives who had collected to watch what was happening.

This form of mimicry is not confined to insects, as we have seen. Even plants take advantage of insects, such as orchid blossoms that mimic female wasps or bees and are pollinated by soon-to-be-disappointed males that try to copulate with them. Wickler also described a particularly interesting and illuminating aggressive mimicry in a marine environment. Small fish known as cleaners stake out territories on coral reefs. Other fish know where to find them and have learned to recognize them by their distinctive appearance and the little "dance" they do to attract customers. Just as we visit our barbers and hairdressers, various kinds of fish show up regularly for a thorough delousing. Customers let the cleaner search their body—even their gills and the inside of the mouth—for external parasites, which it removes and eats. A fish aptly named the saber-toothed blenny mimics the cleaner—even imitating its dance. If a fish looking for a cleaning is hoodwinked by the mimic, it gets a nasty surprise: the mimic takes a big, painful bite out of one of its fins. The startled victim wheels around, "but the mock cleaner calmly stays put as if knowing nothing about it, and remains unmolested because of its cleaner's costume," as Wickler wrote.

A few caterpillars are aggressive mimics that eat other insects. Almost all of the world's 125,000 known species of moths and butterflies (order Lepidoptera) are vegetarians that munch on leaves or other plant parts in the caterpillar stage. Among those exceptions known for many decades are about fifty species that, as Walter Balduf reported, eat scale insects that are immovably attached to a plant or are virtually incapable of moving. Only about thirty years ago, Steven Montgomery made the amazing discovery that fifteen kinds of inchworms, caterpillars of the moth family Geometridae, are ambush predators that capture active, highly mobile insects—unique exceptions among the one thousand plant-feeding species of the worldwide genus *Eupithecia*. These inchworms exist only in Hawaii, where they evolved from a vegetarian ancestor that arrived on these isolated volcanic islands after they had arisen from the sea more than five million years ago. The various species, all suitably camouflaged, wait in ambush on different perches: green leaves or stems, brown twigs or fallen leaves. As other inchworms often do to make themselves inconspicuous, these ambushers use claspers at the posterior end of the abdomen to hold on to their perch, from which they stretch their bodies straight out, and freeze in position. On a twig they look like a short branching spur. "When a prey animal touches [its] posterior end," Montgomery wrote, "the caterpillar suddenly loops backwards to seize the prey with its thoracic legs . . . and quickly returns to a straightened, elevated posture to feed. The entire strike takes about 1/12 second." The thoracic legs are elongated and "armed with enlarged spine-like setae and sharp claws."

Unlike the many web-spinning spiders, only a few terrestrial insects fashion traps to catch their prey. But several years ago, Alain Dejean and his coworkers published an account of traps built by a tree-dwelling ant (*Allomerus decemarticulatus*) of French Guiana. As do quite a few other ants, *Allomerus* has a mutually beneficial association with a particular species of plant. The plant provides leaf pouches in which the ants nest, and the ants reciprocate by destroying insects that feed on the plant. Dejean and his coauthors reported that the ant "uses hair from the host plant's stem, which it cuts and binds together with purpose-grown fungal mycelium [long threadlike strands], to build a spongy 'galleried' platform for trapping much larger insects. Ants beneath the platform reach through . . . holes and immobilize the prey, which is then . . . transported and carved up by a swarm of nestmates."

The most famous of the trap-making insects are the ant lions of the order Neuroptera, relatives of the aphid-eating green lacewings. Harold Bastin described how an ant lion larva digs its pitfall trap in the soil:

The larva is a strange-looking insect, thick-set and somewhat oval in contour, with a flat head armed with formidable, curved mandibles. It has an ingrained habit of walking backwards, and uses its convex abdomen as a plough. When constructing the pitfall for which it is famous, it usually begins by making a circular groove to correspond with the margin of the proposed excavation. It then ploughs round and round in diminishing circles, constantly jerking out the sand with its shovel-like head. The final result is a funnel-shaped hollow, in the bottom of which the maker lies buried with only its ugly jaws exposed to view. Any small insect which chances to run over the edge of the pit slides downward on the yielding sand, its descent being hastened by the ant-lion, which casts up jets of sand upon its victim.

After its fanglike mandibles inject the victim with venom and digestive enzymes that liquefy its internal tissues, the ant lion sucks up its predigested meal—much like a robber fly or an ambush bug.

It is a wonder of evolution that a fly (genus *Vermileo*) maggot, or larva, of the snipe fly family (Vermileonidae) has independently invented this trap. In *Demons of the Dust,* William Morton Wheeler described how this larva, the worm lion, digs its pitfall:

> The procedure is very simple compared with the usual circuitous performance of the ant-lion, because the worm-lion merely curls its anterior end after thrusting it in the sand and then suddenly straightens it, thus tossing the sand out onto the surface. At the same time it rotates more or less on its long axis, so that the direction in which the sand is thrown differs somewhat with each discharge. In this manner a small conical pit, with the larva at its apex, is soon formed . . . when the pit is completed the larva awaits its prey. . . . Usually . . . it lies horizontally on its back with its posterior half buried in the sand and its thoracic and first abdominal segments crossing the floor of the pit like a bar and covered with a very thin layer of sand.

When an insect falls into the pit, the worm lion usually strikes "at the prey violently and repeatedly till it [can] fix its mandibles in some portion of its body." Next "it pump[s] venom into its victim and then commence[s] imbibing its juices."

The delightful robin-sized burrowing owls of the Great Plains and southern Florida are very unusual. Unlike most owls, they are partly diurnal, unusually long-legged, live and nest in burrows that they dig to a depth of as much as 8 feet, and spend much of the day surveying their surroundings from the top of the large mound of excavated soil. If alarmed, perhaps by a birdwatcher or an approaching predator, they give a loud chattering call as they agitatedly—and amusingly—bob and bow.

One of the more remarkable things about these little owls is that they scatter chunks of horse or cow dung (probably bison dung earlier in the species' history) around the entrances of their burrows. The dung is evidently important to them, because if it is removed, they will usually replace it. What purpose the dung serves was a mystery until Douglas Levy and his coworkers showed that it is bait to attract dung beetles, relatives of the scarabs, that the owls eat. The researchers' first step was to remove all dung, beetle scraps, and regurgitated owl pellets (which contain the indigestible parts of beetles and other prey) from around the entrances to twenty occupied burrows. Then they put a quantity of cow dung "typical of the amount [usually] found at a burrow entrance" around the entrances to half of these burrows and none around the entrances to the other half. After four days, they collected the dung beetle scraps and regurgitated pellets from around the entrances to all the burrows. Then they repeated the experiment, switching the bait from one group of burrows to the other. Examination of the beetle remains found on the ground and in the pellets showed that "when dung was present at the burrows, owls consumed ten times more individual dung beetles of six times as many species than when dung was not present." The inescapable conclusion is that burrowing owls use dung as bait. The use of bait to catch insects is, indeed, very unusual for a bird, but not unique. On several occasions a green heron was seen catching fish attracted to bits of bread that it had dropped on the water at the edge of a pond in a park.

Birds have evolved many other truly remarkable anatomical, physiological, and behavioral adaptations for exploiting insects as food. Roger Tory Peterson's words put the birds and insects in an ecological context, neatly setting the stage for a look at birds as predators of insects:

> The insects, which have invaded nearly every terrestrial environment on earth, are unable to evade the birds that probe the soil, turn over the leaf litter, search the bark, dig into the trunks of trees, scrutinize every twig and living leaf. The water is no safe refuge, nor is the air, nor the dark of night. There is a bird of some sort to hunt nearly every insect. Warblers and vireos methodically work the leaves while swallows, swifts and other hunters of flying-insect prey spend most of their waking hours on the wing, ranging hundreds of miles daily in their aerial forays.

This story began about 155 million years ago with the first known bird, the famous *Archaeopteryx,* represented by beautiful, complete fossils from a limestone quarry in Bavaria. *Archaeopteryx* combined avian and reptilian characteristics, which shows—as do more recently discovered feather-bearing dinosaur fossils from China—that the birds are direct descendants of the dinosaurs. During the next 120 million years many species of birds evolved, but relatively few of them were insect eaters. During the Miocene epoch, beginning about 30 million years ago, the rapid evolution of the flowering plants and the hundreds of thousands of insects that exploit them resulted, as Frank Gill noted, in an explosive evolutionary radiation of insectivorous birds, mainly songbirds (order Passeriformes), which today constitute close to six thousand of the almost ten thousand known species of birds.

Most birds include insects in their diet. With few exceptions, most notably pigeons and doves, even the most dedicated vegetarian birds, fruit and seed eaters such as finches, buntings, grosbeaks, and cardinals, feed their nestlings a high-protein diet of animal matter, mainly insects. The behavior of a cardinal observed by Josselyn van Tyne is illustrative: "At noon on May 24 the adult male, on his way back to the nest territory, stopped at my feeding shelf with his beak full of small green worms [caterpillars] such as I had often fed to the young. He immediately put the worms down on the shelf and began cracking and eating sunflower seeds. . . . He then picked up the worms, flew across the street, and (presumably) fed the young."

Insect eaters avoid competition by sharing the environment, specializing in where and how they hunt. Ground feeders such as towhees and fox sparrows search the litter on the forest floor for both insects and seeds; warblers, chickadees, and other leaf gleaners search foliage for caterpillars and other insects; nuthatches and brown creepers are among the bark gleaners; wood and bark probers, such as woodpeckers, bore into trees to find grubs and other burrowing insects; flycatchers and, on occasion, many other birds are air salliers that dash from a perch to snatch insects from the air; and finally, gleaners of aerial plankton, among them swallows, swifts, and nightjars, more or less constantly swoop through the air to scoop up flying insects.

As Peterson elaborated, avian insectivores have become specialized to exploit insects as food in many different ways. Take, for example, the many birds that snatch flying insects on the wing. Most of the thirty or more flycatchers (family Tyrannidae) that nest in the United States and Canada wait quietly on a perch, often a bare branch with a clear view of the surrounding airspace from which they dash out to snap up passing insects. They usually return to the same or a nearby perch to wait for another insect to fly by One of the most familiar of the tyrant flycatchers is the eastern kingbird, often seen perched on fence wires along country roads. This gray and white bird, truly a pugnacious tyrant, attacks and chases away any birds that come too close to its territory, even crows, hawks, and vultures. From Nebraska south to Texas, the exquisitely beautiful pink and pearly gray scissor-tailed flycatcher, its forked tail twice as long as its body, is a "wire bird" that likes to perch on telephone lines along the roadside. The little brown tail-pumping eastern phoebe, which often builds its mud-based nest under a bridge or on a rafter in an outbuilding, is one of the first birds to return to the northeastern United States, as early as March. "Now and then," Edward Forbush and John May wrote, "one of these early birds may be seen darting out from its perch in a March snowstorm, apparently catching insects."

Tyrant flycatchers will eat almost anything that flies, as well as small caterpillars and spiders that "balloon" through the air on long strands of silk. Several species even eat wasps and bees, because they can avoid being stung or can recognize and catch only stingless males. Some bee-keepers believe that the brown-crested flycatcher, which preys on bees in apiaries, is a pest that eats worker bees. A discerning Arizona beekeeper, quoted by Herbert Brandt in *Arizona and Its Bird Life*, at first agreed "but after examining the stomach contents of a large number of [brown-crested flycatchers] during a period of more than 20 years, and not in a single instance finding the remains of a worker bee, nor finding a bee sting in the mouth or throat of one of these birds, became convinced that [they] did not prey on worker bees, but only on drones [the stingless males]."

Many birds are opportunists that will flycatch when a flying insect comes temptingly close. I have seen warblers of several species put their search for caterpillars on hold as they darted out from a leafy branch to catch a passing fly. Even a nuthatch crawling on a tree trunk will stop searching for insects on the bark to pursue a flying insect. Cedar waxwings feed mostly on small fruits such as chokecherries and blackberries. But "in late summer and early fall," Forbush and May noted, "the cedar waxwing turns to flycatching, and taking its post on some tall tree, usually near a pond or river, launches out over water or meadow in pursuit of flying insects." Birds caught at such times have been found crammed to the very beak with insects.

The birds commonly called goatsuckers (order Caprimulgiformes, from the Latin roots for *goat* and *milk*) are seldom-seen night fliers that eat almost nothing but insects. (Their odd common name stems from the European myth that they suck milk from goats.) At night, the loud and clearly enunciated calls of most of the North American goatsuckers not only make their presence known, but tell us their common names, which are verbal renditions of their calls: whip-poor-will, poor-will, and chuck-will's-widow. The common nighthawk is an aberrant goatsucker that flies both at night and during the day and, unlike the others, is likely to live in a city, laying its well-camouflaged eggs not in a nest but on the surface of a flat, gravel-covered roof.

Goatsuckers have wide, gaping mouths with which they scoop insects from the air. Except on the nighthawk, stiff bristles surrounding the mouth expand the "scoop." Goatsuckers eat night-flying insects of almost any kind; among others, mosquitoes, June beetles, flying ants, and moths, including even large moths with 4- to 5-inch wingspans, such as cecropia, luna, and polyphemus. Larger than the other goatsuckers at a length of almost a foot, the chuck-will's-widow occasionally eats small birds. The goatsuckers are the night shift of the birds that harvest the many night-flying insects and ballooning spiders that crowd the air. As we will soon see, they are joined by bats, which fly much higher than most of the goatsuckers.

The swifts, the swallows, and the nighthawks—the last working days only part-time—are the day shift of the birds that subsist on aerial plankton. There is little competition between them because nighthawks are seldom numerous and the swifts and the swallows, which are usually very numerous, have found a way to share the airspace. Swallows tend to hunt a relatively short distance above the ground, usually at a height of only a few yards. Swifts, on the other hand, fly much higher. Chimney swifts, the only members of their family in eastern North America, live in towns and cities and, almost constantly on the wing from sunrise to sunset, fly well above the tallest church steeples. Before the advent of Europeans in the New World, chimney swifts glued their twig nests to the inner walls of hollow trees. Today practically all of them prefer to nest in chimneys.

Several of the eight North American swallows have also become associated with humans. Most of us view the colorful, fork-tailed barn swallow, the most familiar of these birds, with affection. They usually nest inside a barn or other outbuilding, sticking their feather-lined mud nests to a wall or placing them on a rafter or some other support. An opportunistic hunter, a barn swallow takes advantage of every occurrence that makes insects easy pickings. Forbush and May wrote that "it follows the cattle afield or swoops about the house dog as he rushes through the tall grass, and gathers up the flying insects disturbed by his clumsy progress. When the mowing machine takes the field, there is a continual rush of flashing wings over the rattling cutter-bar just where the grass is tumbling to its fall. The Barn Swallow delights to follow everybody and everything that stirs up flying insects—even the rush and roar of that modern juggernaut, the motor-car, has no terrors for it."

Other birds take advantage of similar opportunities. In spring I sometimes see flocks of ring-billed gulls in recently plowed central Illinois fields, harvesting soil-dwelling insects turned up by the plow, probably including fat cutworm caterpillars that would have become night-flying moths; C-shaped white grubs, the larval stage of June beetles; larval click beetles, the skinny, brown wireworms; perhaps the overwintering pupae of a corn earworm moth. In Africa, cattle egrets snatch insects flushed up by large grazing animals. We also see these birds associated with grazing cattle in southern Ontario and most of the continental United States. They were first seen on this side of the Atlantic in northern South America about one hundred years ago, a flock probably aided by the trade winds having crossed the ocean. In tropical America, ant birds (family Formicariidae) follow swarms of army ants, feeding on insects the ants flush up as they advance over the forest floor.

Birds that eat insects associated with trees, as we have seen, can be grouped into several quite different feeding guilds. These specializations are driven by competition between species, which forces birds to share the available insects. In 1934, G.F. Gause pointed out that "as a result of competition, two similar species scarcely ever occupy similar niches, but displace each other in such a manner that each takes possession of certain peculiar kinds of foods and modes of life in which it has an advantage over its competition." This is the competitive exclusion principle.

"Species that coexist in seemingly homogeneous habitats, such as grasslands or spruce forests," Frank Gill wrote, "may segregate their niches even more finely." Five insectivorous wood warblers, colorful species that migrate back and forth from the New World tropics to where they nest in the spruce forests of the north, manage to coexist on the same spruce trees by feeding in different ways on particular parts of the trees, as Robert MacArthur discovered. Gill's concise summary of MacArthur's observations explains how the warblers avoid competing with one another "The yellow-rumped warbler fed mostly in the understory below 3 meters [almost 10 feet], the black-throated green warbler in the middle story, and the blackburnian warbler at the tops of the same spruce trees Sharing the middle part of the trees with the black-throated green warbler, which explored the foliage for food, was the Cape May warbler, which fed on insects attracted to sap on the tree trunk. Sharing the treetops with the blackburnian warbler, which fed on the outer twigs and sallied out after aerial insects, was the bay-breasted warbler, which searched for insects close to the trunk."

In addition to almost all of the fifty species of warblers that can be seen in the United States and Canada, other birds also make a living by searching foliage for insects. A few among the many are the vireos, orioles, tanagers, kinglets, titmice, and chickadees. The spry little black-capped chickadees are agile acrobats that nimbly hop from twig to twig and may hang upside-down as they inspect a leaf for their next meal, which is likely to be a caterpillar. They are particularly interesting because they display what—at least in my view—can only be called intelligence as they search for their prey. Their hunting behavior, Bernd Heinrich and Scott Collins found, is amazingly clever and sophisticated. As we will see in chapter 9, they keep an eye out for partially eaten leaves—those that are tattered or holey—which, they realize, indicate that caterpillars are probably nearby.

Some birds are preoccupied with the trunks and larger branches of trees: bark gleaners, such as the brown creeper and white-breasted nuthatch, and bark and wood probers, such as woodpeckers and, of all things, a very unusual finch on the Galápagos Islands. The brown creeper has an energy-efficient way of searching a tree trunk for insects, spiders, other small creatures, and their eggs tucked away in the crevices of the bark. It begins at the base of the trunk, which it climbs up in a spiral path while conducting its inspection. When it is ready to move on, it spreads its wings and, expending a minimum of energy, glides down to the base of a nearby tree trunk and begins another upward climb as it hunts for food. The white-breasted nuthatch frequently crawls headfirst down the tree trunk. From this perspective it is likely to find food that brown creepers miss. In winter, nuthatches supplement their diet with plant food, such as acorns and sunflower seeds, which they often conceal in bark crevices for future use, a behavior that inspired their common name.

The woodpeckers (twenty or more species in North America), Roger Peterson wrote, "spend most of their lives in a perpendicular stance, clamped against a trunk or a branch, the stiff tail acting as a brace and the deeply curved claws, two forward, two aft on each foot, clutching the rough bark. The straight beak, hard as a chisel, is driven in trip-hammer blows by powerful muscles in the head and neck." The beak is used to find wood-boring insects by gouging into solid wood, and to excavate its own deep nesting cavities. Beetle grubs and other insects are extracted from their burrows by a barbed tongue that can extend as much as five times the length of the bill, a tongue so long that it can be stored in the head only by looping around the skull. A physician quoted by Steve Nadis wondered what makes it possible for these birds to use their head "as a battering ram without sustaining headaches, concussions or other brain injuries," why dead and dying woodpeckers don't litter the countryside. Dissecting woodpecker heads has yielded some answers, among them a tightly fitted skull that keeps the brain from banging around and shock-absorbing muscles that encircle the skull.

The remarkable woodpecker finch is one of a group of fourteen finch species found only on the Galápagos Islands. Discovered by Charles Darwin in 1835, these birds, commonly called Darwin's finches, are very different from one another in feeding behavior and have beaks appropriately adapted to handle what they eat. Among them are species that feed on insects, seeds, leaves, nectar, or the pulp of cactus pads, according to David Lack. It is generally agreed that all of them evolved from a single colonizing flock of one species that somehow crossed 600 miles of the Pacific Ocean from the closest point on the South American mainland to the recently (geologically speaking) volcanically formed and at first lifeless islands. With few other birds to compete with them, they avoided competing with one another by evolving ways of exploiting unoccupied ecological niches.

An ornithologist working on the Galápagos in 1914 was the first to observe woodpecker finches using tools, Lack noted. Although they peck holes into trees to find wood-boring insects, they lack the long extensible barbed tongue with which true woodpeckers extract beetle grubs or other insects from their holes. Instead, as Sabine Tebbich and her colleagues reported, woodpecker finches "use twigs or cactus spines, which they hold in their beaks . . . to push, stab or lever [insects] out of tree holes and crevices. . . . Moreover, they modify these tools by shortening them when they are too long and breaking off twiglets that would prevent insertion."

A few other birds, perhaps two or three dozen species, are known to use tools, but only a handful use them to capture insects. Jeffery Boswall noted three Australian birds—the shrike-tit, the grey shrike thrush, and the orange-winged sittella—that use twigs to probe for insects in crevices. In Tangipahoa Parish in Louisiana, Douglas Morse watched brown-headed nuthatches pry pieces of bark from longleaf pines with flakes of bark to get at hidden insects.

Nathan Emery and Nicola Clayton, in an article in a 2004 issue of *Science,* wrote that wild "New Caledonian crows . . . display extraordinary skills in making and using tools to acquire otherwise unobtainable foods." Tools for extracting insect larvae from holes in trees "are crafted from twigs by trimming and sculpting until a functional hook has been fashioned." Other tools, "consistently made to a standardized pattern" by cutting pieces from *Pandanus* leaves, are used "to probe for [insects] under leaf detritus [with] a series of rapid back-and-forth movements that spear the prey onto the sharpened end or the barbs of the leaf." On foraging expeditions, the crows carry these tools from place to place. One caged New Caledonian crow, Emery and Clayton noted, appeared "to be capable of reasoning by analogy with her previous experience with hooks, by modifying nonfunctional novel material (metal wire) into hook-like shapes to retrieve food."

In winter, cohesive flocks of black-capped chickadees, tufted titmice, nuthatches, brown creepers, golden-crowned kinglets, and downy woodpeckers wander through the woods foraging for insects. The birds, so to speak, gang up on the insects. Field studies summarized by Kimberly Sullivan "showed that individuals can benefit from membership in a flock by decreasing their risk of predation and increasing their foraging efficiency." Flock members constantly sound contact, or social, calls—such as the chickadee's *chick-a-dee-dee*—that announce their presence and help to maintain flock cohesion. They also have calls, such as the chickadee's high-pitched *zeee*, that warn of an approaching hawk or other predator. Most small birds, according to Susan Smith, freeze or dive for cover in response to the warning calls of their own and other species. Sullivan, as we will see in chapter 9, found that downy woodpeckers spend much more time eating and much less time cocking their heads from side to side to watch for predators when they are with a flock than when they are alone, because the constant contact calls of their companions assure them that others are also keeping an eye out for predators.

Many of the 4,500 species of mammals—from tiny shrews, bats, and mice to huge bears—feed on insects to varying extents. Some, such as the African aardvark and the giant anteater of South America, eat nothing but ants and termites. Many omnivores, among them bears, raccoons, opossums, chipmunks, foxes, squirrels, mice, and skunks, include insects in their diets.

Primates such as lemurs, tarsiers, monkeys, baboons, chimpanzees, and humans are omnivores that, to varying degrees, feed on insects. In the early 1960s, Jane Goodall made the famous discovery that chimpanzees create tools from twigs and use them to "fish" for one of their favorite snacks, termites—the tropical species that build large cement-like mounds. Early in the rainy season, swarms of thousands of male and female termites of the reproductive caste leave the mounds through tunnels dug by workers, who keep the exit holes thinly sealed until conditions are favorable for the reproductives to fly off and found new colonies. When a hungry chimpanzee spots one of these lightly sealed holes, Goodall observed, it removes the seal with its index finger and pokes a tool into the hole. A moment later the chimpanzee withdraws the tool and then eats the termites clinging to it. Children in Africa use the same technique to get a few termites for a snack, but adults on the continent make ingenious traps to catch swarming termites—a much-favored food, delicious when roasted—by the thousands. As I explained in *Fireflies, Honey, and Silk,* people of almost all non-Western cultures eat insects, usually as a special treat.

Bats, the masters of the night sky and the only mammals capable of true flight, are not blind, but they find their way in the dark by means of echolocation (sonar). In flight they emit sounds too high-pitched for our hearing and sense obstacles and their prey, usually insects, by listening for the echoes that bounce back from them. As the Nobel laureate Niko Tinbergen observed, on what is to us a quiet summer evening, to the bats flying about and the moths that can hear them "the evening is anything but calm. It is a madhouse of constant shrieking. Each bat sends out a series of screams in short pulses, each lasting less than a hundredth of a second." In chapter 9 we will consider the bat's echolocation in more detail and the question of how moths benefit from an ability to hear bats.

Shrews, which may weigh as little as a tenth of an ounce, are the smallest mammals on earth, and because of their tremendous metabolic rates—their hearts may beat 1,200 times a minute—they are the most voracious of the insectivorous mammals, and probably the most voracious of all mammals. Every twenty-four hours, a shrew eats the equivalent of its own body weight or more in insects, other arthropods, and occasionally a mouse or other small mammal. Shrews live and hunt in extensive runways at or just above ground level.

The mouse-size short-tailed shrew *(Blarina brevicauda)*, common in the eastern half of southern Canada and the United States, is active both day and night throughout the year and is one of the world's few venomous mammals. Delivered in the saliva as the shrew bites, the venom is toxic to both small mammals, which this shrew seldom attacks, and insects, which are the most important part of its diet. The experiments of Irwin Martin showed that crickets and cockroaches are immobilized by the venom but do not die until three to five days after being bitten. "Venom," Martin reasoned, "was therefore acting as a slow poison as well as an immobilizing agent. Immobilization for 3 to 5 days may extend the availability of fresh non-decomposing food, and thus enable *Blarina* to optimally exploit a sudden abundance of insects by caching some. If all hoarded insects were dead, many might [decay and] lose substantial nutritive value before the shrew could eat them."

Insect eaters do, of course, help to prevent insect populations from soaring to ecologically disruptive levels—always a possibility because an insect, depending upon the species, will lay anywhere from a few to thousands of eggs. If, on average, two of a female's eggs survive to become reproducing adults, she will have replaced herself and her mate, and the population of her species will not increase. But if only an additional two survive, the population will increase by a factor of two in each generation and will soon become an ecologically disruptive force. Clearly, dozens or even thousands of a female's offspring must perish—and predators eat many of them.

The many insect-eating animals, from the little crab spiders to birds and even huge bears, consume enormous numbers of insects. In so doing, they exert the powerful selection pressure that results in the evolution of the many ways in which insects can survive by avoiding or defending themselves against predators. A few examples from agriculture show how great the selection pressure from predators can be.

In 1887, sap-sucking cottony cushion scales, insect invaders from Australia, infested California orange groves, threatening to destroy them all. Knowing that these scales were uncommon in Australia, where they were never destructive, Charles V. Riley, a great pioneering entomologist, reasoned that they were controlled in Australia by an enemy absent from California. He postulated that the scale population would crash if this enemy were introduced into California. Therefore a few hundred vedalias, ladybird beetles that eat these scales, were imported from Australia, and in less than two years only a small and inconsequential population of cottony cushion scales survived, coexisting with a few vedalias that kept them in check. In 1945, DDT, which kills vedalias but not the scales, was sprayed in the orchards to control another insect. As was to be expected, seriously destructive outbreaks of cottony cushion scales followed, but the benign balance of vedalias and scales was restored when the use of DDT was discontinued. Robert L. Metcalf and Robert A. Metcalf underscored the importance of predators in controlling pest insects with an example involving two native American insects. In 1899 in Maryland, in just a few days, sieves used in packaging fresh peas separated out twenty-five bushels of hoverfly larvae, which feed on aphids. "They were so abundant that they almost completely destroyed the pea aphids in the fields."

In 1979, Richard Holmes and his coworkers showed that birds alone can significantly decrease populations of some plant-feeding insects. They covered plots of striped maple shrubs in a New Hampshire hardwood forest with nets that excluded birds but not insects. Nearby uncovered areas of similar size and with comparable growths of striped maple shrubs served as controls. The exclusion of birds, especially ovenbirds, black-throated blue warblers, veeries, and Swainson's thrushes, caused a significant increase in the numbers of leaf-eating caterpillars.

Similar experiments by Robert Marquis and Christopher Whelan in Missouri showed that insectivorous birds decreased the number of plant-feeding insects on white oak saplings by half, which in turn allowed the saplings to increase their aboveground growth by one-third. Like Holmes and his coworkers, they covered some saplings with nets that excluded birds and left other saplings uncovered.

Many of you have seen grasshoppers leap into the air and use their wings to make a speedy retreat when you come threateningly close to them. When I turn on the lights in my laboratory at night, panicked cockroaches swiftly run off to find a hiding place. (Entomologists can't use insecticides in their laboratories; insecticides kill not only cockroaches but also the insects that are the subjects of our experiments.) Many insects do not respond to most disturbances by fleeing, because they are camouflaged or hidden, perhaps on the underside of a leaf, under debris on the ground, or in some other nook. Some, however, will leave their hiding place to flee if an intruder comes too close—within a critical distance whose length will vary with the species of the prey insect. The next chapter considers running away and hiding as ways to escape from predators, to avoid becoming a meal for a bird, a mantis, a mouse, or some other insect eater.

# THREE

# Fleeing and Staying under Cover

A well-hidden insect will be safe from many, if not most, insect-eating predators. But since natural selection is inexorable, predators will inevitably evolve with the anatomical and behavioral specializations needed to find and capture even the most thoroughly concealed insects. For example, if you hear what sounds like the blows of an ax in a winter woodland, it may well be a pileated woodpecker, the largest of our surviving North American woodpeckers, using its powerful, chisel-like bill to chop out chips of wood the size of a child's hand as it works to get at the larva of a long-horned beetle hidden deep in the trunk of a tree. An insect burrowing in the soil, such as a wireworm or a white grub, may be found by a mole or the probing bill of a grackle or some other bird. Nevertheless, hiding—although not always successful—can be advantageous, and insects of all sorts, and other animals too, have adopted this strategy for survival.

Natural selection favors—often very much so—an insect's normal lifestyle, especially its feeding behavior, if it keeps the insect out of sight and thereby protected from at least some potential predators. Usually only the larvae, insects in the immature stage, bore into plant tissues, burrow in the soil, or are otherwise hidden, and the usually

Figure 3. Disturbed by a predator, a grasshopper leaps into the air and flies off to make its escape.

immobile pupae generally remain hidden in the larval tunnel or burrow. The much more active adults are exposed to many more predators—spiders, insects, birds, mice, shrews, bats—as they fly and run about searching for nectar or other food, for a mate, and for appropriate places to lay their eggs. Most female insects lay hundreds of eggs, and many are exposed to predators as they fly long distances to distribute their eggs one by one or in small clutches on widely dispersed plants, often of only one or a few closely related species.

In July and August we hear, high in the trees, the loud, shrill drone, the "love call," of male dog-day cicadas—even in cities and towns. The females are frequently on the move as they disperse their eggs in small clutches laid in small cavities slashed into woody twigs by their sharp ovipositors, their egg-laying appendages. After hatching, the tiny nymphs drop to the ground and burrow deep into the soil, where they suck sap from roots until they emerge from the soil as adults about 2 inches long two or more years later. (Because the generations overlap, some cicadas emerge every year.)

Although the nymphs are relatively safe in the soil, the adults are eaten by birds of many kinds. Large, scary-looking but harmless solitary (nonsocial) wasps called cicada killers also search for them in the trees. The wasps inject them with a paralyzing but nonlethal venom, stock each of several small chambers in their underground nests with two or three of them, and lay a single egg in each chamber. (All but a very few nonparasitic wasps feed their larvae insect or spider prey.) The wasp larvae feed on the paralyzed cicadas but remain in the ground—as safely hidden from predators as are cicada nymphs—until they emerge from the soil as adult cicada killers the following summer.

Like the cicada killers, thousands of species of solitary wasps and bees prepare a shelter for their offspring. Most, like cicada killer larvae, live in burrows in the soil, but other parents build aboveground structures that shelter their offspring. (The Nobel laureate Karl von Frisch nicely described and illustrated some of these shelters in *Animal Architecture*.) Some potter wasps (family Eumenidae), for example, build juglike nests of mud that they stock with paralyzed caterpillars, but other wasps of this family are not potters at all and instead nest in hollow plant stems.

Other insects also prepare concealed nurseries to hide their offspring from predators. Carpenter bees (subfamily Xylocopinae), some of which look like large bumblebees, excavate nesting tunnels as much as a foot long in solid wood—on one occasion in the unpainted cedar siding on my house, although they soon gave up because the inch-thick siding was too thin for them. Little brown solitary bees (family Andrenidae) hurry from blossom to blossom in early spring, gathering nectar and pollen from spring beauties. They dig long tunnels in the soil and provision small cells that branch off from the main tunnel with their harvest, which feeds the larvae that hatch from single eggs laid in separate cells. A remarkable mason bee (*Osmia bicolor*) of Europe prepares an individual nest for each of her larvae in the empty shells of land snails, perhaps even those that housed that gourmet's delight the escargot. After finding a shell, stocking it with food, on which she lays only one egg, and blocking the shell's opening, the bee, Frisch explained, "makes a series of flights to collect all kinds of dry stalks, blades of grass, thin twiglets, or even pine needles[;] . . . from this material, she builds a tent-shaped roof over the snail shell, which eventually hides it completely." Like all bees, both solitary and social, she provides her larval offspring with bee bread, a mixture of pollen and honey.

Their activities, seeking mates or places to lay their eggs, make it difficult or impossible for adults to always be hidden. Adult Japanese beetles, June beetles, and other herbivorous relatives of the scarabs feed on the foliage of shrubs and trees. Groups of metallic green and bronze Japanese beetles cluster shoulder to shoulder in conspicuous groups on a leaf. But both of these beetles and related species lay their eggs deep in the soil, including under our lawns. The chubby, C-shaped larvae, known as white grubs, live belowground, feeding on roots. Although well hidden and plagued by far fewer predators than the adults, they are preyed upon by some insects, birds, and moles. Among these predators are wasps of the family Scoliidae, which have no common name. John Henry Comstock noted that these wasps "do not exhibit as much intelligence as do most digger wasps; for they do not build nests and do not transport prey to them for their carnivorous larvae." After locating a white grub in the soil, the female scoliid paralyzes it with a sting, "work[s] out a crude cell about it, and attaches an egg to . . . the grub." The scoliid larva eats the grub, spins a cocoon, and completes its development in its underground cell.

Some immature insects hide in plant matter. The tiny leaf-mining larvae of some beetles, moths, flies, and wasps tunnel in the narrow space between the upper and lower epidermal layers of a leaf, feeding as they go. Their tunnels are clearly visible beneath the translucent epidermis. The tiny apple leaf miner moths glue their eggs to the undersides of leaves. When the larvae hatch, they pass through the egg shell directly into the leaf. Many beetle larvae and moth caterpillars, such as European corn borers, tunnel in the stems of nonwoody plants. Some snout beetles (weevils) gnaw a tunnel into an acorn or other nut with the mandibles at the end of their long, thin snouts and then turn around to place an egg in the tunnel and then move on to lay more eggs. When the full-grown larva emerges from the fallen acorn, it burrows into the soil to pupate. Some fly larvae, such as the apple maggot, and caterpillars, such as codling moth larvae—the infamous worm in the apple—burrow in fleshy fruits, but fly maggots leave the fruit to pupate in the soil, and codling moth caterpillars move away to pupate in a cocoon, often under a flake of bark on a tree trunk.

The sloth moths have what may be the most unusual lifestyle of all the insects, one that keeps them hidden from most, perhaps all, predators throughout the egg, larval, and pupal stages and exposes them only briefly during the adult stage. As adults, the four species of sloth moths, distant relatives of the European corn borer (family Pyralidae), hide in the dense growth of hair on sloths, slow-moving mammals of New World tropical forests that live high in the trees, feeding on foliage. Anywhere from a few to more than a hundred of these little moths may occupy a single sloth.

When sloth moths were first discovered in the nineteenth century, it was supposed that both the adults and the caterpillars lived on sloths and that the caterpillars fed on the plentiful growth of green algae on the sloths' hair or ate the hair itself. But in 1976 Jeffrey Waage and G. Gene Montgomery reported that although they found many adult moths on sloths, they found no eggs, caterpillars, or pupae. But they did find caterpillars feeding on sloth dung. About once a week a sloth descends to the ground to defecate. Hanging from a vine, it scoops out a pit with the long, curved claws on its hind legs, deposits about a cupful of fecal pellets in the pit, and covers it with leaf litter. Female moths briefly leave the sloth to lay their eggs on its yet-to-be-covered dung. The caterpillars eat dung, and when the moths emerge from the pupae in the dung pit, they fly up into the trees to find a sloth. They mate on the sloth, and gravid females leave the animal only long enough to lay their eggs.

Some insects construct their own hiding places. Working together, several hundred newly hatched tent caterpillars (*Malacosoma americanum*) of eastern North America spin a small tent of silk. As the caterpillars grow, they continuously enlarge the tent until it is about 2 feet long. Shaped like upside-down pyramids in the crotches of wild cherry trees, these tents are a common sight along country roads in spring. At night and during the cool parts of the day—early morning and late afternoon—the caterpillars shelter in the tent, where they are protected from many parasites and predators, Terrence Fitzgerald explained in *The Tent Caterpillars*. When it is warm enough, they leave the tent en masse and march nose to tail in single file to a leafy branch to feed, laying down a pheromone trail that will later guide them back to the nest.

Groups of other insects, mainly caterpillars, also cooperate to spin the communal silken nests in which they live. The messy nests of the fall webworm (*Hyphantria cunea*), constructed on the leafy branches of many kinds of trees, are a common sight in late summer in much of southern Canada and the United States. In spring and early summer, the webworm moths emerge from silken cocoons hidden under leaf litter or a flake of bark and lay their eggs in clusters of several hundred on the undersides of leaves. Upon hatching, the caterpillars immediately begin, as Ephraim Felt explained, "to spin a communal web under which they feed. This protecting web is extended to include more and more foliage till finally a considerable portion of a branch may be enclosed." The caterpillars partly skeletonize leaves, eating only the upper surface, leaving the veins and the lower surface intact. "The skeletonized leaves within the nest soon dry, turn brown, and they, with the frass [excrement] and cast skins of the caterpillars, render the nests very unsightly objects."

The caterpillars known as bagworms (family Psychidae) are well named. They live in cocoonlike pouches that they make of silk and decorate with bits of leaves and twigs. The head and thorax can be protruded through an opening in the bag, enabling the caterpillar to crawl and eat leaves. Fecal pellets are expelled through an opening at the other end of the bag. The familiar evergreen bagworm (*Thyridopteryx ephemeraeformis*) feeds mainly on junipers (red cedars) and arborvitae and, like other species of its family, spends virtually its entire life in its bag. The eggs laid by the wingless females overwinter in the bag. In spring, the newly hatched larvae leave the bag and immediately build their own bags, which they continually enlarge as they grow. In autumn, the full-grown caterpillars pupate in their bags. The winged males emerge from the bags but, having vestigial mouthparts and unable to feed, live for only about a day. Drawn by a female's sex-attractant pheromone, a male thrusts his extensible abdomen far up into her bag and inseminates her. Only after laying their eggs do the larvalike adult females—which lack antennae, legs, and wings—emerge from the pupal skin, drop out of the bag, and die.

A cocoon protects many insects, most famously moths, during the pupal stage, when they are especially vulnerable to predators because being virtually immobile, they cannot run away or defend themselves. Before molting to the pupal stage, Comstock noted, caterpillars "make provision for this helpless period by spinning a silken armor about their bodies." As we will see in chapter 9, several of the giant silkworm moths (family Saturniidae) of North America spin very large tough-walled cocoons, in which they spend the winter in the pupal stage. The huge cecropia caterpillar, for example, constructs a double-walled cocoon

3 or more inches long and immovably attached along its length to a sturdy twig.

In 1978, one of the two homes of the Green Revolution, the International Rice Research Institute (IRRI) on the island of Luzon in the Philippines, invited me to visit and develop a method for testing many thousands of rice varieties for resistance to the rice leaf folder, a moth of the family Pyralidae and an important rice pest. The main problem was figuring out a way to raise large numbers of leaf folder caterpillars in the laboratory so that the varietal tests could be done in a greenhouse. If the varieties to be tested are planted outdoors, the results of the test may be inconclusive, because as luck is likely to have it, the natural population of leaf folders will be too small or almost nonexistent that year.

Gottfried Fraenkel, who years before had supervised my PhD research, was invited to do the very same thing with rice leaf folders by the Central Agriculture Research Institute of Sri Lanka. About six months after I arrived at IRRI, Gottfried stopped off to see me on his way to Sri Lanka. When he asked me if I had made any progress, I one-upped my former boss, handing him a copy of a manuscript ready for publication that described my then recently devised method for testing rice plants for resistance to the leaf folder.

In Sri Lanka, Gottfried did other research projects with this insect, including a masterly study published in a Dutch journal in which he and Faheema Fallil wrote of the leaf folder, "Its characteristic behaviour is to spin a rice leaf longitudinally into a roll, by stitching together opposite rims of the leaf, and to feed inside this roll, leaving the epidermis on the outside of the roll intact," as camouflage. Lying aligned with the long axis of the long, narrow leaf, the caterpillar swings the head end of its body from side to side in the same spot as many as one hundred times, forming a thick band of silk fibers that joins the edges of the leaf together. By repeating this procedure as it advances short distances along the leaf, the caterpillar forms as many as thirty such crossbands. "A newly woven band," Fraenkel and Fallil found, "quickly becomes shorter by a process of contraction . . . bringing the rims of the leaf blade closer together. . . . With each succeeding band, this distance becomes shorter until the leaf is completely rolled up."

If an insect's lifestyle does not commit it to living under cover, in hiding—or if it lacks an effective physical or chemical defense—it will most likely have another way of protecting itself against predators. Some insects, as we will see in chapters 4 and 5, are camouflaged, blend in with the background, or resemble an inedible object, but generally speaking, most defenseless species flee to a hiding place when they feel threatened—even camouflaged individuals whose cover is blown.

In 2008, Oswald Schmitz reported that red-legged grasshoppers (*Melanoplus femurrubrum*) respond differently to ambushing "sit and wait" spiders and to "roaming, actively hunting" spiders. Grasshoppers respond to sit and wait spiders, but not to roaming spiders, by shifting from their preferred food plant, a nutritious grass, to goldenrod, which is not a favorite food but on which they are less likely to be killed by a spider.

Most cockroaches, as Thomas Eisner and his coauthors so aptly put it, "crave concealment. Anyone who has shared a kitchen with cockroaches knows that they seek shelter by day and that they are driven to flee for cover at night if a light is turned on." This is the way of not only the tiny minority of cockroaches that have become household pests but also most of the world's almost four thousand other cockroach species, which live in natural settings.

An insect that hides in a crevice or under a fallen leaf, a flake of bark, a rock, or a clod of soil would, ideally, have eyes not only on its head but also on its tail end so that it could tell if all of its body was safely tucked away in the dark of its hiding place. No insect or other arthropod has eyes on its tail end, but according to M.S. Bruno and D. Kennedy, a crayfish, a spiny lobster, and a shrimp have what Sir Vincent Wigglesworth called a "dermal light sense" at the tail end of the abdomen; in other words, in the "skin," or exoskeleton (chitinous body wall). Actually it is not the skin but some part of the nervous system below it that can sense light. The American cockroach, and probably other cockroaches and many other insects, Harold Ball reported, has a similar light sensor at the tail end of the abdomen. There, a ganglion—a bundle of nerve cells and a part of the ventral nerve cord, which is, roughly speaking, the equivalent of our spinal cord—perceives light that passes through the translucent skin above it. Ball and other researchers demonstrated the existence of a dermal light sense. The American cockroach and other insects can tell light from dark even if the eyes on their heads have been covered with black paint.

Some insects, like cockroaches, rush to a hiding place if sufficiently alarmed. Others, such as the houseflies you startle in your kitchen, fly away rapidly but alight in plain sight on another wall. In either case, the fleeing insect was probably well served by an early warning system. "For species that are palatable," Malcolm Edmunds pointed out, "it will be of advantage if they can detect their predators before the predators detect them, and if they can initiate their active defence (flight) before, or as soon as possible after, the predator has noticed them."

Early warnings may be perceived by the organs of touch, vision, or hearing. At the tail end of the abdomen, noted R. F. Chapman, insects with gradual metamorphosis, except the true bugs, bear a pair of antenna-like sensory appendages, tactile receptors called cerci. Each cercus is clothed with fine hairs ultrasensitive to air currents or touch. This is, of course, an early warning system that alerts the insect to the approach from behind of something that might be a threat. Kenneth Roeder, an entomologist and neurophysiologist, described how the early warning signal of the cerci can be triggered. He suggested that "at night when cockroaches are most active, the observer should slowly approach a single insect standing motionless near the center of an unobstructed area such as a wall or floor. A short puff of air directed at the cerci . . . will send the roach scurrying off and probably out of sight."

The early warning signal, a nerve impulse, travels along the insect's ventral nerve cord from the cerci to the ganglion in the thorax that controls the legs and launches the insect on its escape to safety. The faster the warning signal travels, the better. Among the many nerve fibers that constitute the ventral nerve cord of some insects, including the cockroach, are six to eight giant fibers as much as fourteen times as thick as the others. The virtue of the giant fibers is that they conduct nerve impulses much more rapidly, according to Roeder, at a rate of close to 23 feet per second rather than the other fibers' rate of no more than 2 feet per second.

Their ability to perceive distant objects often makes the eyes the most effective of the early warning systems. Most adult and nymphal insects have two compound eyes on the head, and many also have simple eyes (ocelli) between the compound eyes. Larvae have only simple eyes. The unusual structure of an insect's compound eye—radically different from that of our eyes or those of other vertebrates—gives it an exceptional ability to sense movements. A compound eye is an aggregation of closely packed but separate light-sensitive elements. "The system of small units . . . which constitute the compound eye," Chapman explained, "lends itself to the perception of changes in stimulation resulting from small movements of [an] object." This translates into a highly sensitive early warning system. For example, if you've ever tried to snatch a sitting fly, you know that it will, unless you are very fast, dash away long before your hand can reach it.

Except in the case of certain singing insects, Robert and Janice Matthews observed, "one does not usually think of insects as possessing ears." Male singing insects—cicadas, crickets, and katydids are among the most familiar—belt out "love songs" to attract a mate. Females obviously must have ears, but males have them too, so they can listen for competing males. Most insects that don't make sounds do not have ears. But some mute insects of several unrelated groups—moths, lacewings, praying mantises—do have ears. If they are not listening to each other, what are they listening to? The answer, Roeder demonstrated, is fierce predators: night-flying, insect-eating bats.

In chapter 2 we saw that bats find their way in the dark by means of echolocation, a discovery made by Donald Griffin. They sense obstacles and flying insects by using their keen sense of hearing—many have very large ears—to listen for the echoes of their own sounds, pitched too high for us to hear, that bounce back from these objects. Flying insects that hear bat cries, Roeder found, respond by taking evasive action, which differs with the species: power-diving, folding the wings and falling to the ground, changing course, flying faster and more erratically.

An insect may flee from a predator by running, jumping, swimming, or flying, but many, notably plant feeders, just drop to the ground. With few exceptions, the insect is most likely to survive if it drops as soon as possible. The shaking of the branches and leaves of a tree or shrub may signal the arrival of a predator, most likely a bird hopping from twig to twig as it scans the foliage for insects. In response to a disturbance of this sort, some insects immediately bail out by dropping to the ground, a disappearing act that may happen even before the bird notices the insect. Some insects, notably aphids, disturbed by a bird— or more likely a ladybird beetle or an aphid lion—may simply drop from the plant. But, as Malcolm Edmunds notes, "few aphids respond to a predator by dropping, [although] this is always a successful method of escape. One disadvantage of dropping is that the animal may have great difficulty in finding and climbing a suitable plant on which to feed, particularly if it is immature and has no wings." This is not a problem for leaf-feeding caterpillars of many species—and some spiders—which lower themselves more carefully and may climb back up on a thin strand of silk.

Some beetles, particularly snout beetles, respond to a disturbance by tucking in their legs and letting themselves tumble to the ground, where they are likely to lie motionless for some time. Many, as seen in the beautiful color photographs in Stephen Marshall's encyclopedic *Insects: Their Natural History and Diversity*, look deceptively like dark fecal pellets or small clods of earth. Among the latter group is the well-disguised quarter-inch-long plum curculio (*Conotrachelus nenuphar*), which is brown with a few small white markings and four large humps on its back (wing covers). Fruit growers take advantage of these insects' escape behavior to find out if they are numerous enough in plum, peach, or apple orchards to justify the expense of an insecticide application. "Jarring the beetles from the trees in the early morning, on sheets placed on the ground," explained Robert L. Metcalf and Robert A. Metcalf, "enables the grower to . . [census the population] of these beetles."

The forked fungus beetles (*Bolitotherus*), named for a pair of prominent "horns" that protrude from just behind the male's head, have an escape behavior similar to but even more impressive than the plum curculio's. Adults and larvae of this eastern North American species feed on the shelf, or bracket, fungi commonly seen protruding from the trunks of dead trees. If they are disturbed, Marshall observed, the adults' "first defensive response is to pull in their appendages and drop to the ground." Legs and antennae protectively retract into special grooves. The beetles don't move, "feigning death," and are difficult to spot because they look even more like clods of earth than the plum curculio. In addition, adult forked fungus beetles have a chemical defense, an irritating substance secreted by an eversible forked gland at the tip of the abdomen. But the most amazing thing about these beetles is their early warning system, described by Jeffrey Conner and his coworkers. The beetles recognize "mammalian breath on the basis of its temperature, moisture, and flow dynamics," which causes them to evert the gland, but they do not respond to a mechanically produced air stream. The beetle's "ability to cue in on mammalian breath enables it to respond preemptively to a potentially lethal attack from a ground-foraging mammal, perhaps a deer mouse. Gland eversion can save the beetle by making it distasteful at the very moment that it is taken into the predator's mouth, before a bite is inflicted."

Many insects, especially ground-dwelling species such as cockroaches and beetles, run away as rapidly as they can when alerted by their early warning system. There are, however, surprisingly few records of how fast they can go. My friend and colleague Fred Delcomyn, a neuro-physiologist and an expert on the neural control of walking and running by insects, told me that it is very difficult to time running insects because they rarely go very far in a straight line. The American cockroach, according to G.M. Hughes and P.J. Mill, is one of the fastest insects, running at a maximum speed of 51 inches per second, or 2.9 miles per hour. This may seem slow, but consider the rate compared to body length. The American cockroach runs a distance of about thirty-four times its body length of 1.5 inches in one second. This is equivalent to a coyote with a body length (excluding the tail) of 2.8 feet running 65 miles per hour. I doubt coyotes can run that fast. (I once clocked a *panicked* one going 45 miles per hour trying to outdistance my pursuing car as it ran along a ditch bank parallel to the road I was driving on.) It seems, then, that on the basis of a fair comparison, the American cockroach is probably a faster runner than a panicked coyote. Keep in mind that some of the thousands of cockroach species that live outdoors—and surely have many enemies other than irate householders—run at least as fast as and perhaps even faster than the pestiferous American cockroach.

A sudden leap into the air is another good way to escape from a predator. It is not surprising, then, that many different kinds of insects, members of several unrelated groups, have evolved this escape tactic independently of one another—and have jumping apparatuses fashioned in quite different ways, and even from different body structures. Grasshoppers, crickets, and katydids are closely related and probably inherited their jumping hind legs from a common ancestor. But fleas, flea beetles, and certain relatives of the aphids and cicadas, such as leafhoppers and planthoppers, have leaping hind legs that obviously evolved independently of each other and may be quite different in design. Click beetles and springtails, close relatives of the insects that I discuss below, leap into the air without using their legs.

Grasshoppers are the jumping insects familiar to most people. Their front and middle legs are of the usual walking type, but the hind legs are obviously modified for jumping. The tibia, the longest and thinnest of the leg segments—equivalent to our lower leg—articulates on the femur, equivalent to our thigh. The femur, the most massive segment of the leg, is greatly swollen to accommodate the powerful muscles that move the tibia. A grasshopper preparing to jump raises the front part of its body on its walking legs and flexes the joints between its hind femurs and tibias. It launches itself into the air as the brawny muscles in the femurs suddenly and with great force extend the tibias, straightening the legs as the tibias thrust against the ground to propel the grasshopper as much as a foot into the air and a linear distance of nearly 30 inches, about fifteen times the length of its body.

Adult fleas, the champion jumpers among the insects, use their remarkable ability mainly to board the host animals from which they suck blood. They also leap to escape enemies, most likely the host itself, perhaps a scratching dog, less likely a flea-eating predator. A flea's hind legs are marvelously adapted for leaping but function differently than those of a grasshopper. The powerful femoral muscles jam the coxa, the basal section of the hind legs, against a pad of a protein called resilin in the coxa's socket, thereby greatly compressing it. (There is more on resilin later in this chapter, when we consider insect wings.) When a trigger mechanism releases the "cocked" hind legs, "the expanding resilin," as Howard Evans so concisely explained, "drives the legs downward, and the powerful leg muscles send [the flea] leaping through the air, end over end but landing head forward."

The plant-feeding click beetles of the family Elateridae are among the few insects or insect relatives that can hurl themselves up into the air without using their legs. Comstock said of them, "There is hardly a country child that has not been entertained by the acrobatic performances of the . . . click beetles, or skip-jacks. Touch one of them and it at once curls up its legs, and drops as if shot; it usually lands on its back, and lies there for a time as if dead. Suddenly there is a click, and the insect pops up into the air several inches. If it comes down on its back, it tries again and again until it succeeds in striking on its feet, and then it runs off." The beetle, lying on its back, arches its body so that it is supported only at one point at its front end and another at its hind end. It then tenses the muscles that will straighten its body. When a triggerlike click mechanism abruptly releases this tension, the beetle's body suddenly straightens with a click, slamming into the ground and catapulting the insect into the air.

Click beetles, observed Walter Linsenmaier, "usually make use of their unique ability only when fleeing danger. In a typical situation the beetle is on a plant when threatened. It immediately drops to the ground and feigns death. Then suddenly it snaps itself into the air." The leap itself is likely to startle a predator. And as Harold Bastin noted, conspicuous eyespots on the upper side of the thorax of some click beetles, such as the North American eyed elators, which lie motionless with legs and antennae tucked out of sight if they land back side up, give them "the appearance of villainous-looking little reptiles—a resemblance which probably serves to scare away their enemies." When Eisner and his colleagues offered click beetles to orb-weaving spiders, "the spiders invariably attacked but the beetles commenced clicking the moment they were seized, and in many cases succeeded in freeing themselves. Experiments with wolf spiders gave similar results, as did tests with a mouse and a jay, although with these two vertebrates the elaterids escaped only rarely."

The springtails, until recently considered to be insects, have been removed from class Insecta to a class of their own, Collembola. These tiny arthropods, plant feeders or scavengers, live mainly in the soil. Their jumping mechanism is unique, consisting of a furca shaped like a two-tined fork and hinged at the rear end of the abdomen. The furca can be swung forward under the abdomen and held tautly flexed against it by a catch called the retinaculum. The furca strains against the reti-naculum under considerable tension. When it is abruptly released, the furca swings down and backward forcefully, propelling the springtail high into the air and away from any predator that may have disturbed it.

Red-spotted purples drink nectar, but, like one I was stalking with my insect net in hand, they also sip fluids from dung—a taste shared with some other butterflies. Although seemingly engrossed, my quarry sensed a slight movement and, before I could swing my net, sprang into the air and flew off. Taking to the air and flying away as swiftly as pos-sible is certainly one of the most expeditious ways to escape from a threatening predator, and the great majority of insects have functional wings in the adult stage, although never in the nymphal or larval stages. "In some insects," R.F. Chapman noted, "special types of flight are associated with this escape reaction." For example, the grasshoppers we flush as we walk through a weedy field abruptly leap into the air, fly a short distance, and then make a precipitous crash landing. Escaping moths of several species, Philip Callahan found, fly erratically, making frequent sharp turns and steep high-speed ascents. He observed swift, day-flying white-lined sphinx moths (*Hyles lineata*), often mistaken for hummingbirds, easily escaping from red-headed woodpeckers, mock-ingbirds, and loggerhead shrikes "by evasive action consisting of sharp climbs and turns." Callahan watched a red-headed woodpecker give up the chase after attempting to follow a white-lined sphinx in a steep climb of about 200 feet.

How fast can insects fly? The ground speed depends on the spe-cies and varies with environmental conditions. For example, Chapman pointed out that "if the insect is moving downwind its ground-speed may exceed its air-speed, but in upwind orientation the converse must be true." He also said that "flight activity is broadly divided into two categories, trivial flight, which is concerned with feeding and mating, and migration, during which these vegetative activities are suppressed so that flight behaviour predominates." He did not mention the third and very important category, flights to flee from predators, in which the escape velocity is likely to be considerably faster than the cruising speed.

"A large dragonfly," according to Howard Evans, "can attain a cruising speed of 18 mph and a honey bee about 14 mph." Sir Vincent Wigglesworth summarized the known ground speeds, most likely cruising speeds, of several flying insects: a green lacewing, 1.3 mph; the white cabbage butterfly, 5.6 mph; a bumblebee, 11 mph; a horsefly, 31 mph; and a sphinx moth, 33.5 mph. Unfortunately, data on the escape speeds of insects are scarce. The top speed, presumably the escape speed, of a large Australian dragonfly is about 36 mph, according to May Berenbaum. Chapman cited a field observation showing that a migratory locust (a kind of grasshopper) flies fastest at takeoff and then slows to a cruising velocity of about half the takeoff speed. This suggests that the escape velocity of a frightened grasshopper, and probably of many other insects, is likely to be considerably faster than its cruising speed.

The question of how insect and bird flight speeds compare is ecologically significant because birds are the most ubiquitous and, especially during the nesting season, the most voracious daytime vertebrate predators of insects. Most birds, Frank Gill found, fly at speeds between 19 and 37 mph, and Roger Tory Peterson pointed out that small birds, the majority of avian insectivores, seldom exceed 30 mph. Clearly, most birds are faster fliers than all but the fastest insects. Furthermore, a bird's *attack speed* may be much faster than its cruising speed. I know of no data for insect-eating birds, but hummingbirds, Gill noted, fly faster than usual to "beat competitors to nectar-filled flowers." Insects compensate for their slowness, at least to some extent, by timely alerts from their early warning systems and by their erratic, dodging flight.

The smaller the bird, the greater the wing-beat frequency. The great egret, a large bird more than a yard long, flaps its wings 2 times per second, Peterson noted; the 10-inch-long mockingbird, 14 times per second; and the tiny, 3-inch-long ruby-throated hummingbird, a nearly maximal 53 beats per second. According to Wigglesworth, it is the same with insects. A saturniid moth with a wingspan of 4 to 5 inches beats its wings about 8 times per second; the much smaller honeybees and houseflies, 190 times per second; the even smaller mosquitoes, 600 times per second; and a really tiny midge with a wingspan of only eight-hundredths of an inch makes an astonishing 1,047 beats per second.

For many years there were two hypotheses to explain how insects manage to beat their wings with almost unbelievable rapidity. One proposed that the wing muscles are awesomely (and implausibly) efficient. The other, shown to be correct by Michael Dickinson and John Lighten, is that the downstroke of the wing muscles is augmented with power stored by "springs" that stretch as they brake the wings' upstroke. The springs are made of resilin, mentioned above, probably the most elastic substance known, with an elastic efficiency of 97 percent, meaning that only 3 percent of the energy required to stretch it is lost as heat.

When I sit in my recliner and look out my window, I often see cottontail rabbits on the lawn. If someone approaches, the cottontails do not immediately scamper away. They freeze in place, not moving a muscle, until the intruder comes threateningly close. Only then do they run, having first relied on their camouflaging coat of brown and gray to keep them hidden in plain sight. (They don't seem to realize that brown is conspicuous against a background of green grass.) Like the cottontail and other small mammals, most insects—probably the great majority—may not, at least at first, be noticed by birds and other predators because, as we will see in the next chapter, they blend in with their background, be it foliage, sand, or leaf litter.

# FOUR

# Hiding in Plain Sight

My first and most impressive experience with a camouflaged animal—still indelibly etched on my brain—happened more than sixty years ago. It was an encounter with a snake rather than an insect, but it is a convincing example of how deceptive camouflage can be. Sam Silver, one of my teachers, and I stood on a rocky slope at the base of a tall cliff in Trumbull, Connecticut, our binoculars focused on a wood thrush. My left foot rested on a dished-out rock covered with the usual forest floor litter, mainly dead leaves of various shades of brown and reddish brown. I felt a tap on my boot and looked down to find out what was going on, but I saw nothing. Another tap, and I still saw nothing. After a third tap, I examined the rock again, and there was a venomous copperhead, which, fortunately, was too small to strike above the top of my boot. (Later I did find drops of venom on my boot.) I hadn't immediately seen the snake; it was an all-but-perfect match for the dead leaves. But then, as if by magic, the snake became visible, seeming to rise up from its background. It reminded me of a common optical illusion, a two-dimensional grid of shapes that appear concave but after you do a "mental twist" seem to rise up from the paper.

Figure 4. Many creatures have face patterns, often black lines, that tend to mask the eye—as do the raccoon, insects like leaf-hoppers, and the loggerhead shrike.

Camouflage, or crypsis, is the first line of defense of many, if not most, small animals. You have probably noticed that most small mammals, including mice, rabbits, squirrels, woodchucks, and muskrats, are an inconspicuous brown or gray. (The blatantly conspicuous black-and-white skunks are exceptions—for good reason.) As you approach a rabbit, it freezes, relying on its camouflage to keep it safe, but when you come too close, it scampers off as fast as it can, the usually concealed white underside of its short tail flashing. And so it is with insects. Caterpillars, katydids, and most other insects that feed on leaves are green. During the day, many night-flying moths rest motionless on the bark of a tree trunk, their camouflaged wings making them virtually invisible. Inchworm caterpillars munch on leaves during the night but in daylight are very difficult to spot, as they, without making a move, pose as spurs, short branches of twigs.

Some nectar-sipping insects are colored and patterned so as to be practically invisible on bright, multicolor blossoms. A small moth (*Schinia masoni*) nectaring on the daisylike yellow and red blossoms of *Gaillardia* in the western United States, as Lincoln and Jane Van Zandt Brower observed, usually rests so that its yellow head and thorax are over the yellow inner area of the flower's central disk and its red front wings, which cover the dark brown hind wings, are over the wide, red outer margin of the central disk. The Browers observed that seventeen of twenty moths rested in this position, which made them inconspicuous, rather than in some position in which their color pattern did not match the flower's pattern.

"The very existence of camouflaged grasshoppers, mantids, pupae, caterpillars, and many other animals," Denis Owen argued, "indicates intense selection by predators: if there were no selection animals would not harmonize with their background." Moreover, evidence from experiments and systematic observations, summarized by Hugh Cott and Malcolm Edmunds, shows that being cryptic does make it less likely that an animal will be found and eaten by a predator. But keep in mind that neither camouflage nor any other form of protection is perfect. There will always be some individuals that become meals for hungry predators.

One of the earliest experiments on insect camouflage was done in Italy in 1904 by A. P. Di Cesnola of Oxford University with the large European mantises (*Mantis religiosa*), which have since become established in North America. This insect occurs in two color forms, green and brown. "It is interesting to notice," Di Cesnola wrote, "that the green form is always found upon green grass, the brown form upon grass burnt by the sun." He tethered twenty green mantises on green plants and another twenty-five on brown plants, and twenty brown ones on brown plants and forty-five on green plants. He checked them daily from August 15 to September 1. All of the mantises that matched the color of the plants to which they were tethered survived. "Of the 25 green individuals exposed on brown grass, the last was killed . . . eleven days after the commencement of the experiment. Of the 45 brown individuals on green grass, only ten were left alive on September 1st." That evening a gale swept away the remaining mantises. Birds had killed nearly all the mantises that had died.

Pupae of the black swallowtail butterfly (*Papilio polyxenes*) come in two color forms. Those of the fall generations, which do not become adults until the following spring, are always brown and difficult to find among the dead plants of winter. Caterpillars of summer generations become brown pupae if they molt to the pupal stage on a tree trunk or other dark surface but become green if they pupate on a green plant. The pupa's color, Wade Hazel and his coauthors explained, is controlled by a "browning" hormone produced by neurosecretory cells in the larva's thorax shortly before it molts. If the hormone is released, the pupa will be brown; if not, it will be green. Experiments done by Hazel's group show that the pupae's camouflage gives them significant protection from predators that hunt in daylight. The scientists glued both brown and green pupae to either matching or contrasting natural backgrounds outdoors. Green pupae were significantly more likely to survive on a green background and brown pupae on a brown background.

Others have done similar experiments. In Texas, F. C. Isely tethered grasshoppers of four different colors on sod or soil backgrounds that either corresponded to or contrasted with their coloration. More than a hundred wild birds of several species found and ate 84 percent of the grasshoppers that were conspicuous because they contrasted with their background, but the birds found only 34 percent of those that were difficult to spot because they blended in with the background.

In the Netherlands, L. De Ruiter presented hand-raised European jays that had never seen an inchworm with freshly killed inchworms, also called stick caterpillars, randomly scattered on the floor of a cage among similar-looking short lengths of dead twigs. One of seven jays found its first caterpillar after only one minute, but the other six took much longer, an average of twenty-four minutes. "Many... birds will, if they have accidentally found ... a caterpillar," De Ruiter observed, "go on pecking at every similar object afterwards. ... The bird, therefore, is likely to find nothing but real sticks for some time to come." This soon discourages the bird from pecking at either sticks or stick caterpillars.

One of my graduate students, Aubrey Scarbrough, had an experience that demonstrates the survival value of camouflage. More than a dozen mutant blue larvae appeared among the very large green cecropia caterpillars he was rearing under a net covering a small apple tree. A vandal with a BB gun, probably a mischievous boy, shot all but three of the conspicuous blue mutants right through the net, but not one of the more numerous, well-camouflaged green caterpillars. Birds, not boys, are the usual agents of selection that weed out poorly camouflaged insects, but both boys and birds have color vision and hunt by eye. Years later I found John Gerould's 1921 article on a similar mutation in caterpillars of a butterfly, the clouded sulphur. He found that blue mutants were soon seen by birds and eaten but the cryptic green caterpillars were seldom noticed.

The effect of air pollution on the peppered moth (*Biston betularia*) in England and other countries, known as industrial melanism, is a geographically widespread, unintentionally created experiment. Its results prove that camouflage works and are also a compelling demonstration of how natural selection acts, in this case, to change the moths' color. Before the Industrial Revolution, these moths, which were pale and peppered with tiny black spots, were inconspicuous when they spent the day resting motionless on the bark of tree trunks covered with a growth of light-colored lichens. The Industrial Revolution began in England with the invention of the steam engine in the late eighteenth century. Coal-burning factories proliferated. In woodlands near factory towns, smoke blackened the tree trunks and killed the lichens that grew on the bark. The pale moths were now conspicuous on the dark bark. The first black moth was found near Manchester in 1848. By 1895, according to Bruce Grant and L.L. Wiseman, about 98 percent of the peppered moths near Manchester were black, and this once-rare color form spread to other areas blackened by industrial soot, and flourished. In unpolluted areas, pale moths remained the predominant form.

H.B.D. Kettlewell of Oxford University did the pioneering studies of industrial melanism. He proposed and eventually demonstrated that in smoke-polluted areas, birds that scan tree trunks for insects are the selective agents that eliminated pale moths and usually spared the black form. First he showed that peppered moths placed on tree trunks of a matching color, either dark or light, were much more likely to survive until evening than those placed on tree trunks of the "wrong color—determined by trapping the survivors after they flew off following nightfall." The next step was to find out what had happened to the missing moths. By watching from blinds, he saw that birds caught the moths, often finding those that did not match the color of the bark on which they had been placed but rarely those that had been placed on bark of a matching color.

After the passage of the United Kingdom's Clean Air Act of 1956, air pollution diminished and lichens again grew on trees near factory towns, once again favoring the pale form of the peppered moth. Amazingly to some but predictably, the course of natural selection rapidly reversed when, because of the change in the moths' environment, being black was no longer advantageous. Grant, Denis Owen, and C.A. Clarke reported that "beginning in 1959 the [peppered moth] population at Caldy common, 18 km west of Liverpool, has been sampled each year. There the frequency of the [black] form has dropped from a high of 94.2% in 1960 to its current (1994) high of only 18.7%.... Similar reversals are well documented elsewhere in Britain."

Cryptic color patterns and the behavioral adaptations that make them effective are more complex than you might think. In his 1957 *Adaptive Coloration in Animals,* Hugh Cott explained that "in its crudest and most generalized condition a cryptic pattern serves to break up an animal's form into a number of more or less contrasted patches of colour whose shapes are arbitrary. While the shapes of these patches entirely fail to suggest to the eye the form of the body on which they are superimposed, they do not necessarily suggest anything else in particular." Then Cott made the crucial observation that "a further step towards invisibility is taken when the disruptive design more or less closely resembles the particular environment against which it is normally seen." (Also see below.)

   This means that a camouflaged animal not only needs to recognize and orient to an appropriate background but must also assume a body posture and a behavior that will, as Cott emphasized, "bring the [camouflaging] pattern into appropriate relation to the environment simulated." Many night-flying moths, for example, spend the daylight hours napping quietly on the bark of a tree, as does the peppered moth. Generally speaking, these moths spread and press their wings flat against the bark, largely eliminating the telltale shadows that would reveal their three-dimensional shape. Many moths bear stripes or other linear markings that may extend across the breadth or the length of their wings. Most moths orient their bodies so that these markings align with the "vertical cracks and shadows of the bark," in Cott's words. Markings that extend across the breadth of the wings will be properly aligned if the moth's body is in a vertical position. Those that extend along the length of the wing will be in alignment if the body is at a right angle to the vertical. As we saw above, the position assumed by *Schinia masoni* on *Gaillardia* blossoms is a perfect example of "correct" orientation.

Not long before sunrise on a day in August, an underwing moth (genus *Catocala,* family Noctuidae) with a wingspan of more than 2 inches lands on the trunk of a white birch and covers its hind wings with its front wings. It disappears! Well, it doesn't really disappear; it's just that its white front wings, marked with a few black streaks, blend in so well with the white bark of the birch that the moth is all but invisible. "If you were to peel off a wedge of bark from a white birch," John Himmelman wrote, "you would have the twin of this moth." As he pointed out, the resting moth is so well camouflaged that an insect-eating bird or other hungry predator is not likely to spot it. The moth will remain motionless as it clings to the birch's bark until it flies off in the dark of the coming night. Sometime shortly before sunrise on the next day, the moth—guided by its instincts—will settle down on the trunk of one white birch or another. Oddly enough, leaves of the white birch are not among the preferred foods of this moth in its caterpillar stage. The caterpillars regularly feed on aspen and poplar and, according to Frank Lutz, only occasionally on white birch.

This moth's close relatives, all fellow members of the genus *Catocala,* use exactly the same strategy to avoid being discovered in daylight by an insectivorous bird searching for a meal. But because they rest on the trunks of other kinds of trees—with dark bark—their front wings are camouflaged with subtle patterns of brown or gray that are usually marked with zigzag lines. However, the hind wings of most of the *Catocalas* are blatantly conspicuous, black in ground color and, depending upon the species, crossed by broad bands of bright yellow, orange, or red—except for *Catocala relicta,* the one that spends the daylight hours clinging to the trunks of white birches. Its hind wings are black, but each has a broad band of contrasting bright white. The *Catocalas* are commonly known as underwings because of their often colorful hind wings, which are completely hidden by the cryptic front wings when the moth is at rest. In his book on these moths, *Legion of Night,* Theodore Sargent noted that the word *Catocala* is derived from the Greek roots *kato,* "behind," and *kalos,* "beautiful." He wrote, "'Beautiful

behinds'—perhaps I missed the best title for [my] book!" The conspicuousness of the hind wings is important because, as we will see in the
next chapter, it might save the moth's life if its first line of defense, its
camouflage, fails.

The white underwing is Himmelman's favorite of the more than one
hundred species of *Catocala* that live in the United States and southern
Canada. He has seen this more northern moth only twice in the vicinity of his hometown, Killingworth, Connecticut. I like to think that he
found them while "sugaring" in the woods at night. Sugaring, a collecting technique favored by moth aficionados, including my friend Jim
Sternburg and me, consists of swabbing a sweet, fermenting bait on the
trunks of trees shortly before sunset and then returning at night with a
flashlight to look for tipsy moths that have come to sip the intoxicating
bait. (On sunny days, I have seen red-spotted purples, large, beautiful butterflies, sipping from fermenting fruit under an apple tree—so
inebriated that they staggered and so stupefied that they could be
picked up with the fingers.) Collectors, Sargent reported, have sought
to "devise ever more irresistible 'brews' . . . and many . . . swear by their
own, often secret recipes." He himself uses a simple mixture of brown
sugar and stale beer. My favorite bait is made by combining molasses,
sugar, canned peaches, and stale beer and letting the mixture ferment
for a day or two. William Holland wrote about his sugaring experience
in 1903:

> Here we have a bucket and a clean whitewash brush. We have put into
> the bucket four pounds of cheap sugar. Now we will pour in a bottle of
> stale beer and a little rum. . . . Before the darkness falls, while yet there
> is light enough to see our way along the path, we will pass from tree to
> tree and apply the brush with the sweet semi-intoxicating mixture to the
> trunks of the trees. . . . It is growing darker. . . . Now let us light our [kero-
> sene] lamps . . . [and] retrace our steps along the path and visit each moist-
> ened spot upon the tree-trunks. . . . Just above the moistened patch upon
> the bark is a great *Catocala*. The gray upper wings are spread, revealing
> the lower wings gloriously banded with black and crimson. In the yellow
> light of the lantern the wings appear even more brilliant than they do in

sunlight. How the eyes glow like spots of fire ... Let us go to the next tree. ... Here [the moths] are holding a general assembly. Look! See them fairly swarming about the spot. ... There is a specimen of *Catocala relicta,* the hind wings white, banded with black. How beautiful simple colors are when set in sharp contrast and arranged in graceful lines!

*Catocalas* spend the winter season in the egg stage, as do many other insects. In summer, Sargent noted, the moths lay their eggs "on the trunks of trees, either singly or in small clusters, and usually tucked into crevices in the bark." Until the eggs hatch the following summer, about nine months later, they are prey for bark-gleaning birds such as nuthatches, brown creepers, and downy woodpeckers. But because of their small size, gray color, and rough texture, they are difficult for the birds to spot, and quite a few survive.

Like their parents, *Catocala* caterpillars are active at night, feeding on foliage, and immobile during the day. When they have grown large, caterpillars of most species of *Catocala* stretch lengthwise vertically as they cling to the bark of a twig, branch, or tree trunk and are thus incredibly well camouflaged. Their color is a close match for the color of the bark, and markings that resemble the scars left by shed leaves or fallen twigs enhance the deception. In many species, a fringe of short filaments on each side of the body dangles down to the bark, obscuring the caterpillar's shadow and the space between its body and the bark. Full-grown caterpillars descend to the ground and wrap themselves in fallen leaves bound together with silk before they molt to the pupal stage. During the three or four weeks before the adult moth emerges, their flimsy shelter surely conceals them from some predators but probably not from others, such as shrews and white-footed mice.

Animals that sleep at night and are up and about during the day have camouflage-enhancing characteristics seldom or never seen in animals that sleep during the day and are active at night: masking of the eyes, disruptive coloration, countershading, deceptive movements, and modifications of its surroundings.

The characteristic targetlike shape of a vertebrate's eye, with its round, staring black pupil, is likely to draw the attention of a predator to even the most deceptively camouflaged fish, frog, snake, bird, or mammal. Hugh Cott's book gives many examples of how black lines or a mask like the racoon's make the eyes of vertebrates less conspicuous. Insects' eyes lack a pupil and are usually but not always less conspicuous than those of vertebrates. But a grasshopper from Brazil is, according to Cott, "a green insect whose brown eyes strongly contrast with the general body colour[,] ... a feature likely to attract attention. ... [But when] we examine the grasshopper among broken surroundings, ... the eyes are scarcely noticeable[,] ... obliterated by two broad brown stripes ... each [of which] runs from [the eye] backwards across the head, extending unbroken over the thorax onto the wings." It looks, Cott described, as if an artist drawing the insect "had accidentally smudged the eye while still wet and drawn a trail of ink or paint right along his figure."

Many animals, Denis Owen pointed out, "are marked with bold stripes or bands which are either much darker or much lighter than the rest of the coloration. Such markings are called disruptive coloration and serve to break up an otherwise distinctive and characteristic body outline." Disruptive markings help to camouflage many animals, among them fish, frogs, birds, mammals, and many insects. Tobacco hornworm caterpillars, the big green "worms" that eat the leaves of tomato plants, have a row of white disruptive lines on each side of the body. Long black lines that cross both wings obscure the form of the tiger swallowtail butterfly. The most famous of the disruptively marked butterflies, at least among North American entomologists and butterfly observers, is the white admiral, a member of the same species (*Limenitis arthemis*) as the red-spotted purple, which is not disruptively marked. Disruptive stripes are characteristic of almost all of the twenty or more species of the genus *Limenitis,* as well as of other genera of the Nymphalidae, such as some species of tropical *Anartia.*

The wings of the white admiral, a resident of the north woods area of southern Canada and the northern United States, are, as Jim Sternburg, Arthur Ghent, and I described, traversed by broad white bands that contrast with the otherwise dark wings and extend from the leading edge of the front wing to the trailing edge of the hind wing. When this butterfly is seen out of its natural context—as a specimen impaled on a pin in a box, for instance—the white bands are glaringly obvious, and it is by no means apparent that they are camouflage. However, when I collected white admirals on Michigan's Upper Peninsula, I found that they were not easy to spot, especially when they sat among small rocks while sipping water from the moist soil at the edge of a dirt road. In Illinois I can see a red-spotted purple from a distance of many yards, but when white admirals sat among rocks in Michigan, I usually did not see them until I had come so close that they would flush before I could net them. As we will see in chapter 10, the red-spotted purple, found south of the white admiral's range, has abandoned its disruptive coloration because it has evolved a very different way of warding off predators.

As recently as 2005, Innes Cuthill and his coauthors were able to say that there has been only one experiment in nature to test the hypothesis that disruptive coloration tends to protect an animal from predators. Robert Silberglied and his coauthors on Barro Colorado Island, Panama, published the results of that experiment in 1980. They obliterated the white bands, much like those of the white admiral, on the blackish upper surfaces of the wings of the butterfly *Anartia fatima* with a black felt pen and put similar control marks on the adjacent dark areas of the wings of a control group, leaving their appearance unchanged. The scientists repeatedly captured and released these butterflies for twenty-one weeks and found that the experimental and control groups did not differ significantly in longevity or wing damage. They interpreted this to mean that these butterflies' disruptive patterns did not protect them from predators that hunt by sight.

However, Jim Sternburg and I pointed out that blackening the white stripes produced a butterfly that resembled several distasteful black butterflies in that area. Therefore, the edible *Anartias* with blackened disruptive bands were artificially created mimics protected from predators by their resemblance to the distasteful butterflies. The experiment is inconclusive because there is no objective way to choose between the Silberglied group's and our interpretation of the data.

To this day, there are no unambiguous demonstrations of the efficacy of an animal's disruptive coloration in nature. However, the Cuthill group's field experiments with artificial models support two key predictions of the hypothesis: "that patterns on the body's edge should be more effective than equivalent patterns placed randomly" and "that highly contrasting colours should be more disruptive than those of low contrast." (This is how science works. Hypotheses are tested by determining if predictions derived from them are correct.) The models, colored paper triangles with dead mealworms (succulent beetle larvae) pinned to them, were pinned to the bark of oak trees in a nature reserve in England. The ground color of the triangles matched that of the bark. As the hypothesis predicts, models with high-contrast disruptive markings that extended to the edges of the triangle were less likely to be noticed by birds than were other models with low-contrast markings. Statistically, there is only one chance in a thousand that this difference could have resulted from random chance.

"When an animal or any other solid object is observed out of doors in the open," explained Cott, "it will be seen that its upper surface is more brightly illuminated than its under parts, owing to the direction of incident light from the sky. The effect of this top lighting is to lighten the tone of the upper parts, while the lower surfaces, which are shaded by the body, appear to be darkened." It is this contrast of light and shade that reveals to us, even in a two-dimensional photograph, that an object, say a cylinder or a ball, is actually three-dimensional. No matter how well its coloration blends in with its background, a camouflaged animal that is obviously three-dimensional will soon be noticed by a bird or other predator.

You have probably noticed that almost all animals—fish, snakes, birds, rabbits, insects—are dark on their upper surfaces and light on their lower surfaces, an arrangement called countershading. If an animal's upper and lower surfaces were of the same color, the lower surface, shaded by the upper part of the body, would appear to be darker than the better-illuminated upper surface, and the contrast would reveal the animal's three-dimensional form. But if the lower surface is of an appropriate light shade that blends seamlessly with the darker upper surface, the revealing contrast is eliminated. Cott's words are as clear as, or clearer than, anything I could write: "The artist, by the skillful use of light and shade, creates upon a flat surface the illusionary appearance of roundness: nature, on the other hand, by the precise use of countershading, produces upon a rounded surface the illusionary appearance of flatness."

Some caterpillars are dark above and light below, but others, among them relatives of the tobacco hornworm, are reverse countershaded, light above and dark below. The reason for this is, of course, that the reverse-countershaded caterpillars normally rest upside-down, clinging lengthwise to the underside of a stem. Cott's book has photographs of a caterpillar of the eyed hawk moth taken outdoors. When the caterpillar is in its normal upside-down position on the underside of a twig or stem, it is inconspicuous because there is absolutely no contrast between its upper and lower surfaces. But when Cott posed it backside-up on the upper side of a twig, it stood out like a sore thumb because of the eye-catching contrast between its upper and lower surfaces.

Camouflaged animals, such as the rabbit frozen in place, the moth sleeping on the trunk of a tree, or the mantis waiting to ambush its next meal, usually blow their cover if they move. But sometimes movement enhances camouflage. True to their name, the dead leaf butterflies of India hide from birds and other predators by resting among dead leaves clinging to a branch. They and various other leaflike butterflies will gently sway from side to side to simulate movement in the breeze. As

I write, hanging on the wall before me are two framed specimens of the Asian dead leaf butterfly. The upper surfaces of the spread wings of one seem conspicuous and do not look at all like dead leaves. The other, perched on a dead twig with its wings held together in the resting position, could easily pass for a dead leaf. A pattern of lines on the grayish undersides of its wings suggests the veins and midrib of a leaf. Short, closely overlapping tails on its hind wing look like a leaf stem, and curved, overlapping projections at the tips of its front wings suggest the "drip tip" so characteristic of the leaves of tropical trees.

Major R. W. G. Hingston, who led Oxford University's 1929 expedition to British Guiana (now Guyana), described an immature mantis's camouflage-enhancing movements in *A Naturalist in the Guiana Forest*. The mantis waits quietly for an unsuspecting insect to come close as it clings, head down, to the bark of a tree. Except for its abdomen, its body is gray, matching the bark's color, and marked with patches of green and yellow that simulate lichens that grow on the bark. Although the rest of its body remains motionless, its gray-green abdomen, which droops down over its back and resembles a leaf, sways so as to mimic the trembling of a leaf in the breeze, "sometimes gently, at other times vigorously, just as a leaf is made to move when touched at one time by a puff of air and at another by a distinct breeze."

A few early naturalists described insects modifying their surroundings to enhance their camouflage. Cott mentioned "a remarkable [green] South American caterpillar . . . which gnaws the leaf of its food plant in such a manner as to leave uneaten a number of rough models of itself . . . attached to the mid-rib. It then takes its station on the mid-rib beside them." Lost in the crowd, so to speak, the motionless caterpillar is not easily located. Major R. W. G. Hingston observed a caterpillar in British Guiana that arranges a similar deception in a different way. The caterpillar feeds on a plant with long leaves, baring about a third of the outer end of the midrib. It then cuts off some pieces of about its own size from the leaf blade and attaches them to the bare part of the midrib with threads of silk. When the caterpillar, of about the same color as the pieces of leaf blade, perches at the tip of the midrib, it is perfectly concealed.

In his beautifully illustrated book *Insects of the World,* Walter Linsenmaier noted that "many young insects, especially young bugs and beetles, disguise themselves ... by bestrewing themselves with sand, soil, or even their own excrement. They may be studded with hooks or bristles that help to hold the materials." Mark Moffett noted that certain unusual ants (*Basiceros singularis*) he observed in Ecuador "are uncommonly dirty, camouflaging themselves with mud held in place on their bodies by feathery hairs." Compared to scurrying run-of-the-mill ants, these exceptionally slow-moving ants are all but invisible. They move at a snail's pace—not a problem if your favorite prey is, in fact, snails.

One of the assassin bugs (family Reduviidae) is known as the masked bedbug hunter because, like some other members of its family, it disguises itself by covering its body with debris. The "nymphs," Stephen Marshall reported, "exude a sticky substance that soon accumulates a body-masking layer of junk ranging from the dead bodies of victims to dismembered dust bunnies. Next time you spot some wandering lint in the corner of your living room, take a closer look to see if it's a Masked Bed Bug Hunter nymph."

In one form of camouflage, usually more descriptively known as special resemblance or mimesis, the animal does not try to disappear by blending in with its background. It stays in plain sight and is likely to be conspicuous, but it is disguised to look like something inedible, something of no interest to a hungry predator, such as a bird dropping, a twig, or—in association with others of its species—even a spikelike inflorescence of small blossoms.

# FIVE

# Bird Dropping Mimicry and Other Disguises

The caterpillars of some of our common swallowtail butterflies are disguised as bird droppings, as are some other insects. For example, larvae of the familiar black swallowtail, called parsleyworms when they invade our gardens, have a special resemblance to bird droppings in their first two instars (subdivisions, separated by molts, of the caterpillar or any other larval stage) but are camouflaged in their last two instars.

In 1892 in India, Colonel A. Newnham was reaching across a bush to collect a beetle and, as he said, "nearly touched what I took to be the disgusting excreta of a crow. Then to my astonishment I saw it was a caterpillar half-hanging, half-lying limply down a leaf." The caterpillar, disguised as a bird dropping, was a very deceptive and convincing imitation of its model. Newnham was struck by "the skill with which the colouring rendered the varying surfaces, the dried portion at the top, then the main portion, moist, viscid, soft, and the glistening globule at the end. A skilled artist working with all materials at his command could not have done it better."

H. O. Forbes discovered an interesting spider in Java when he seized a butterfly (a skipper) that seemed to be resting on a bird dropping on a leaf. To his surprise, part of the body remained behind, adhering to the dropping. When Forbes touched the "dropping," he found that his eyes

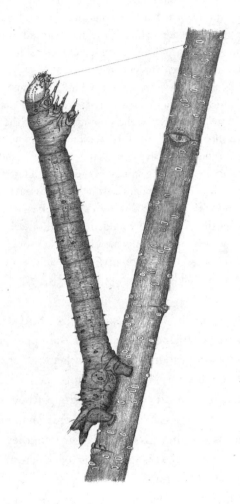

Figure 5. During the day, this nocturnal inchworm caterpillar remains motionless, disguised as a twig on a tree, a strand of silk bracing its difficult to maintain posture.

"had been most perfectly deceived, and that the excreta was a most artfully coloured spider" setting a trap for its prey. He commented:

The appearance of the excreta rather recently left on a leaf by a bird or a lizard is well known. Its central and denser portion is of a pure white chalk-like colour, streaked here and there with black, and surrounded by a thin border of the dried-up more fluid part, which, as the leaf is rarely horizontal, often runs for a little way towards the margin. The spider . . . is of a general white colour; the underside, which is the one exposed, is pure chalk white, while the lower portions of its first and second pair of legs and a spot on the head and on the abdomen are jet black.

This species does not weave a web of the ordinary kind, but constructs on the surface of some prominent dark green leaf only an irregularly shaped film of the finest texture, drawn out towards the sloping margin of the leaf into a narrow streak with a slightly thickened termination. The spider then takes its place on its back on the irregular patch I have described, holding itself in position by means of several strong spines on the upper sides of the thighs of its anterior pairs of legs thrust under the film, and crosses its legs over its thorax. Thus resting with its white abdomen and black legs as the central and dark portions of the excreta, surrounded by its thin web-film representing the marginal watery portion become dry, even to some of it trickling off and arrested in a thickened extremity such as an evaporated drop would leave, it waits with confidence for its prey."

Caterpillars of a certain moth in Africa also look like bird droppings. They rest on the leaves of their food plant—on the upper side, of course. (Feces dropped by a bird flying over or perched higher in a tree can never land on the underside of a leaf.) When they are young and small, these caterpillars are gregarious, and in groups they give the impression of many little droppings excreted by several small birds roosting higher in the tree. When they grow to a larger size, they are no longer gregarious, and they spread well apart. Then each caterpillar looks like a single dropping of a large bird.

R. W. G. Hingston described a moth that usually spends the day resting on the upper side of a leaf, its wings spread wide and flattened down against the leaf. The wings are a glossy, almost pearly white with a slaty gray blush, marked with dark brown patches and traces of yellow. Lying in its usual position, the moth looks exactly like a small bird's dropping that has fallen from a height and flattened against a leaf.

A leaf beetle (*Neochlamisus platani*) that riddles the leaves of sycamores and plane trees with small holes has evolved a similar tactic to protect itself from insect eaters such as birds The adult beetle looks like the fecal pellet of a large caterpillar, a convincing resemblance because of its black color, peculiar shape, rough and bumpy upper side, and appropriately small size—only about 0.18 inch long. The females camouflage their egg masses by covering them with their own droppings, and in the larval stage, these beetles live within baglike cases composed of their own fecal pellets.

Some foliage-feeding insects of three unrelated groups have independently evolved different ways of achieving an uncanny resemblance to twigs growing on living plants: walking sticks, or phasmids (order Phasmatodea); grasshoppers (order Orthoptera); and caterpillars and moths (order Lepidoptera).

Although most of our North American walking sticks have long, narrow, sticklike bodies and pass themselves off as twigs, some tropical species look like leaves. The twiglike walking sticks generally have very long, almost thread-thin legs and move slowly and deliberately. Their long legs, explained Malcolm Edmunds, enable "them to rock gently from side to side or to 'teeter' backwards and forwards. In this way they can move slowly but mask the movement since the rocking or teetering resembles a twig swaying in the wind." Another walking stick (*Parasosibia parva*) freezes in a position that makes it virtually indistinguishable from a short stub jutting out from a twig. As Cott observed, this insect rests motionless on a branch with its head downward and its front legs and long antennae stretched downward in a straight line and appressed to the twig as tightly as possible. The second pair of legs clasps the supporting twig. The body behind the head "is inclined outward [and upward] at an angle, rigid and straight, with the legs closely applied to the body."

Cott also described an Australian grasshopper (*Zabrochilus australus*) that creates the same deception in quite a different way. It rests head down with its body tightly appressed to the supporting stem and its "fore feet and antennae ... lying together and being closely applied to the stem—precisely as with the Phasmid." The middle legs grasp the stem, and the hind legs stretch straight up, held as close to the stem as possible, "but here the comparison ends." The "twig" angling up from the stem is not the grasshopper's body but its tegmina, the narrow, leathery front wings, "these being so constructed that they cannot be closed down against the body," as can the front wings of other grasshoppers.

The twiglike caterpillars of the family Geometridae (the peppered moth is a Geometrid) are commonly known as loopers, inchworms, or measuring worms because of their peculiar form of locomotion, described by John Henry Comstock: "They progress by a series of looping movements. They first cling to the supporting twig or leaf by their [front] thoracic legs; then arch up the back while they bring forward the hinder part of the body and seize the support . . . near the thoracic legs, by the prolegs at the [hind] end of the body; then letting loose the thoracic legs . . . they stretch the body forward, thus making a step; this process is then repeated." Inchworms have the usual three pairs of legs on the thorax, the three-segmented body region just behind the head, but while most other caterpillars have five pairs of stubby, fleshy prolegs on the abdomen, inchworms have only two pairs, one on each of the two hindmost segments of the abdomen. The family name is from the Greek word meaning "a measurer of land." Cott thought these caterpillars "undoubtedly some of the most perfect known examples of special resemblance." The likeness of an inchworm to a twig of its food plant is often extraordinarily precise. The head is shaped to suggest the end of a twig, and the connection between the prolegs and the twig is obscured by fleshy tubercles that, because they are light in color, "also tend to neutralize the shadow which might otherwise betray the junction." The body may bear lumps that suggest a bud or a short stub, and marks that look like leaf scars.

Most inchworms, Harold Bastin explained, feed at night, "but when daylight comes they take a firm hold upon a twig with their . . . prolegs and stretch out [headfirst] at an acute angle. To lessen the strain that this posture imposes on the body, many of the species spin a delicate silken thread from their mouth down to the stem on which they rest; and that considerable reliance is placed upon this support may be judged from the fact that if the thread be severed the creature falls back with a jerk."

Each species of inchworm, reported Denis Owen, "resembles the twigs of its own food plant, and if placed on a different plant the effectiveness of the camouflage is much reduced." The host plant is not only an insect's food but also the place where it lives. As is to be expected, natural selection has honed insects' camouflage and special resemblances to conform to the often distinctive colors and structures of their habitual host plants. Keep in mind that most of the more than four hundred thousand plant-feeding insects are host-plant specific, picky eaters that will feed on only a very limited number of the more than three hundred thousand known species of plants. Indeed, many will feed on only a handful of closely related species.

An adult moth, the black-blotched *Schizura* of the family Notodontidae, mimics a short branch of a dead twig, as seen in a beautiful photograph in John Himmelman's *Discovering Moths*. Standing on its head like one of the walking sticks described above, with its narrow wings tightly appressed to its body, it sticks up from its perch on a dead twig at an acute angle. This moth's wings are mainly dark gray, but their tips are light gray and resemble loose flakes of bark.

Wherever lichens are a conspicuous feature of the environment, as on tree trunks or stone walls, a variety of insects and other animals have a remarkably deceptive resemblance to them. (Lichens, although they may look like a pale moss, are actually cooperating colonies of a fungus and a green alga.) The irregular dimensional form of a lichen is mimicked not by the animal's shape but, as Cott described, by "the most ingenious and deceptive disruptive patterns, which give the optical impression of irregular processes and deep interstices—even when painted, as they often are, on the flat canvas of a moth's wing or on the ovoid abdomen of a spider."

Lichenlike animals occur all over the world: on sand flats in Illinois, on a limestone wall in England, on rocks almost anywhere in the mountains, and on the bark of trees in South American rain forests. These animals are a diverse group—including, in addition to insects and spiders, lizards and tree frogs, such as the gray tree frog of the United States and southeastern Canada. Among the insects and their relatives are lichen-mimicking daddy longlegs, spiders, mantises, stick insects, grasshoppers, moths and caterpillars, weevils, and long-horned beetles.

The protective value of looking like a lichen is illustrated by Cott's experience with a marbled beauty moth (*Bryophila perla*) that was resting on a lichen-covered wall. Remember that birds search for insects visually, just as we do.

> Well do I remember one occasion when, intending to record its cryptic likeness, I had arranged my camera tripod some twelve inches in front of a Marbled Beauty that had settled on an old wall near Bradford-on-Avon.... I looked up before making the exposure to be sure that the moth was still in position—when I discovered that I had evidently disturbed it while making the final adjustments to the camera. At any rate, the moth had vanished. To make sure of this, however, I subjected the stone where it had been resting to a close and most careful examination, verifying the exact position with reference to the axis of the camera, but I could see no sign of the moth. Just as I had begun to convince myself that I had lost my subject and was about to turn and replace the dark slide, I suddenly *recognized* what I must repeatedly have *overlooked*. My *Bryophila* was there all the time: it had never moved, and was "staring me in the face" in full view.

One of the most astonishing examples of special resemblance—an especially deceptive one—is created by a cooperating group of twenty or more African planthoppers (genus *Ityraea*) of the order Homoptera. About a quarter of an inch long and rather broad winged, these insects look a bit like small moths. There are two color forms of both males and females, green winged and pink winged.

In his 1896 *The Great Rift Valley,* J. W. Gregory of the British Museum (Natural History), now the Natural History Museum, described a startling discovery he and his companion had made in East Africa: "I was working through the woods beside the Kibwezi River with Mr. Watson, one of the missionaries at the station there, when my attention was attracted by a large brightly-coloured flower, like a foxglove." This "flower" appeared to be a spike of many small blossoms and buds ascending a vertical stem, much like the inflorescence of a verbena. But when he touched the spike, Gregory was startled because "the flowers and buds jumped off in all directions." He pointed out a similar "inflorescence" to Mr. Watson, also an amateur botanist, who tried to pick it and was as surprised by the absconding blossoms and buds as Gregory had been.

The arrangement of the planthoppers on the stem, Gregory noted, "with the green bud-like form at the top of the stem, and the pink flower-like insects below, looked so much like an inflorescence" that it was virtually indistinguishable from the real thing. Gregory's observation and colored drawing, the frontispiece of his book, have, as Cott remarked, attracted much interest and some criticism, particularly of the obviously improbable depiction of the green bud-like planthoppers gradually decreasing in size toward the top of the "inflorescence," as buds do in real spikelike inflorescences, where the youngest and smallest are at the top. "I am indebted to Sir John Graham Kerr," Cott noted, "for the information that Gregory subsequently told him the illustration was erroneous as regards the difference in *size* of the individual insects; but in other respects Gregory stood by the correctness of the figure." Gregory's report has been confirmed by similar observations of the cooperative crypsis of groups of planthoppers in various parts of the world, including one unpublished observation near Urbana, Illinois, by Jim Sternburg and me.

# SIX

# Flash Colors and Eyespots

No defensive tactic is perfect. Even the most deceptively camouflaged insect may be spotted by a hungry bird. Some, as we have seen, beat a speedy retreat by running or flying away when a predator comes too close for comfort. Among them are the underwing moths (*Catocala*), which we met in chapter 4. These moths, like many other insects, augment their retreat with another tactic. Before fleeing, an underwing disturbed by an experimenter or a keen-eyed bird, perhaps a noisy gray jay, the "camp robber" of the north woods, suddenly raises its front wings to expose its broadly banded, vividly colored, attention-grabbing hind wings. This startling display of color may serve the moth in two ways. First, the bird may be frightened away or at least startled, which will give the insect precious extra seconds in which to make its escape. Second, if it does pursue the moth, the bird may be deceived into not noticing the moth after the insect lands on the trunk of another tree. The predator is most likely to focus on the flashing colors of the hind wings, strikingly conspicuous as the moth flies away, but these wings suddenly disappear when the moth lands on the trunk of another tree and again covers them with its camouflaged front wings. The bird, which has a search image of something brightly colored, may not notice the now-inconspicuous moth. Similarly, I remember hitting

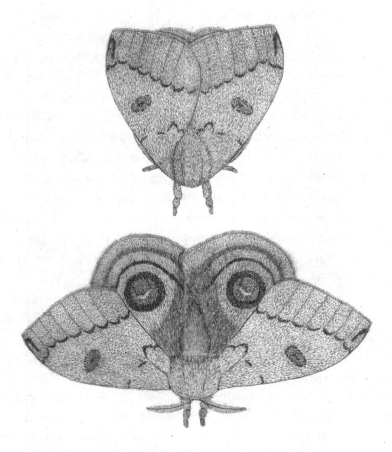

Figure 6. The io moth, active only at night, is motionless in daylight, depending on its camouflage to hide it from insect eaters. If discovered and disturbed, perhaps by a bird, it raises its front wings to reveal two scary eyespots.

a red-banded croquet ball into the rough and taking quite a while to spot it. It was lying in plain sight but the red band was not visible, and I had a search image of red.

Many other insects, some other arthropods, and even a number of frogs, lizards, birds, and other vertebrates have evolved functionally equivalent fallback defenses that, as Denis Owen argued, may deter predators or at least distract them long enough to make escape possible. Some insects, like the underwing moth, abruptly expose vivid hues, called flash colors, while others display eyespots, which may be realistic representations—complete with a highlight on the pupil—of the eye of a bird, lizard, or other vertebrate predator. Visual distractions like flash colors and eyespots are usually referred to as startle displays. Animals may also use their defenses to protect their offspring instead of themselves. In response to the approach of a possible nest robber, many ground-nesting birds such as killdeer and other plovers move away from the nest and do a broken-wing act. This display is a distraction that presumably entices the intruder, sometimes a birdwatcher but more often a predator, to pursue the apparently disabled and easily captured adult.

A defensive distraction need not be visual. Sounds or chemical emissions can serve the same purpose. For example, a cicada grasped in the fingers makes a loud, shrill squawk that, under other circumstances, might startle a cicada-eating bird, causing it to release the victim held in its beak. Another fallback defense is a chemical distraction, the emission of an irritating, foul-smelling, or foul-tasting secretion. The dark brown squash bug, one of the true bugs and a garden pest that sucks sap from squash and pumpkin vines, is well camouflaged as it lurks on the ground in the shade of its host plant. If squeezed, it emits a repulsive odor from glands on its back.

Many insect groups other than underwings and quite a few other moths include some species that display flash colors as a defensive distraction: grasshoppers, walking sticks, mantises, true bugs, cicadas, planthoppers, butterflies.

Most grasshoppers depend upon their remarkably powerful jumping hind legs to bound away from birds and other insect eaters. In a single leap, according to Peter Farb, a large grasshopper may cover a distance of 30 inches. Some grasshoppers also rely on flash colors, Hugh Cott noted, with brightly colored membranous hind wings that are usually folded like a fan out of sight beneath the narrow, leathery, camouflaged front wings, the tegmina. The hind wings of a southern European species "are bright crimson with a blackish border," and those of "a fine green species from the Amazon . . . are purple." In case it can't get away fast enough, the legs of some species are armed with formidable sharp spines, with which the grasshopper will attempt to slash any predator that grabs it.

As a young budding naturalist roaming through the beautiful Connecticut countryside (much of it now covered by housing developments) on a fine summer day, I was frequently surprised when I flushed a large brown grasshopper, a Carolina grasshopper (*Dissosteira carolina*), along a roadside or in a dry field. Its grayish brown coloration blends in with that of the ground, where it frequently sits. Relying on its camouflage, it freezes in place until a bird, a person, or some other threatening presence comes too close for comfort. Then it leaps into the air and flies off, making a crackling sound and with yellow-bordered black hind wings flashing conspicuously. After landing on the ground several yards away, it quickly covers its colorful hind wings with its grayish brown front wings, turns to face the intruder, and remains completely motionless, its camouflage making it almost invisible. A closely related species, the band-winged grasshopper of California and southern Oregon, behaves in much the same way, but each of its hind wings is marked with a large bright red patch.

In a report on the 1899–1900 Skeat expedition to the Malay Peninsula, Nelson Annandale described an unusual defensive display by a grasshopper. "When the hooded locust is taken in the hand it makes very little resistance," probably because its jumping hind legs are "less highly developed" and not as useful in defense as those of other grasshoppers. Instead, it everts a large, "vividly scarlet" bladder (no doubt by means of blood pressure), which otherwise remains hidden, inverted behind the head. The everted bladder projects behind the head like a hood, making the grasshopper quite conspicuous. "To a human observer it appears that he has injured his specimen, and that some brilliantly coloured portion of its internal anatomy is issuing from its neck."

A flash color will not be very startling unless it is hidden in one way or another before abruptly coming into view, only visible when the insect (or other animal) feels immediately threatened by an intruder that might be a predator. This is beautifully illustrated by the startle-display configuration of an Amazonian forest-dwelling cicada. As Cott pointed out, both wings of most cicadas are totally transparent and colorless from base to tip. But the bases of this Amazonian cicada's hind wings are "decorated with a vivid splash of vermilion which extends nearly half-way along the wing." Cott explained that "this color would, of course, be perfectly visible through the transparent fore-wings in the resting insect, were it not for the fact that a patch of about equal area on these wings is pigmented with opaque olive-green, so that when the wings are folded the cryptic green area closes like a shutter so as just to hide the red areas on the wings beneath." Thus, when it is revealed, the red patch on the hind wings comes as a startling surprise.

Except for one in Florida, the forty-four or so North American stick insect species are completely wingless, but many of the approximately 2,500 others, which are mainly tropical, do have wings. There is, however, an evolutionary tendency for these insects to become wingless or, probably more often, for their wings—usually the front ones—to get smaller. Even so, "stick insects," Malcolm Edmunds stated, "often display by erecting the [front] wings, thus giving an apparent increase in size and revealing bright colours or eyespots on the hind wings." As Cott noted, when the front wings are extremely small or altogether absent, as in a species from the Malay Peninsula, the first part of its large and flightworthy hind wing, which lies uppermost when the wing is folded, is cryptically colored and hides the membranous, colorful part of the wing, which is folded like a fan beneath it. In the New Hebrides, both wings of a flightless stick insect (*Cnipsus*) are reduced to small remnants that function as nothing more than a defensive display. When a threatening predator sees through this insect's camouflage and approaches too closely or attacks, the little cryptic front wings raise to display the small, vividly red hind wings.

The io moth (*Automeris io*), one of the wild silk moths (family Saturniidae), is common throughout the United States and the eastern half of southern Canada. It is well equipped to startle a predator, such as a bird, or even a person. It keeps busy at night but spends the daylight hours resting on vegetation, often a leaf, with its cryptic front wings covering its hind wings. If the moth is disturbed, it abruptly spreads its front wings, revealing the colorfully banded hind wings, each dominated by a large eyespot, complete with a highlight on the pupil, that is a convincing imitation of the eye of a vertebrate. As a matter of fact, with its wings spread it looks rather like the face of a small owl. In reminiscing about his dawning interest in insects during his boyhood in Uruguay, Thomas Eisner, the world's foremost expert on the chemical defenses insects use against predators, recalled his first experience with a related and similar moth, *Automeris coresus*: "I remember being startled when I first came upon an *Automeris* and wondered whether predators might be similarly faked out by the eyes."

Another North American wild silk moth (*Antheraea polyphemus*) has similar eyespots on the hind wings and performs a similar distraction display. If that isn't enough to scare away the predator, the moth, described the late John Bouseman and James Sternburg, increases its efforts to be conspicuous by dropping to the ground and bouncing "along . . . by downward thrusts of the forewings [while] elevating the hindwings at each bounce." Similarly, a very large katydid of the New World tropics is camouflaged as a dead, diseased leaf. When disturbed, it spreads its wings, holding the leaflike front wings to the side and revealing the colorful hind wings with large eyespots near their tips.

A Venezuelan praying mantis, like a few other mantises, respond to birds with an amazingly complex display that even some people have described as alarming. Héctor Maldonado described the elements of this elaborate display: the mantis lays back its antennae and opens its mouthparts wide, revealing the colored mandibles; raises the long first segment of its thorax and extends and flexes the pair of grasping legs there to the side, showing a large black spot on each femur (the first long segment of the leg); displays two large eyespots by raising the front wings and raising and spreading the shiny, patterned hind wings; twists its abdomen to the side, revealing otherwise invisible colored bands; and finally makes a rustling noise by stridulating (rubbing body parts together) while sidling and swaying from side to side.

By putting a bird into a cage with one of these mantises, Maldonado tested the efficacy of this insect's elaborate display. He used several birds of each of four species but, unfortunately, did not say how many. Canaries and Java sparrows, seed eaters, elicited a display, although they did not attack the mantis and even stayed well away from it. Orioles and shiny cowbirds, insect eaters, also elicited a display. Shiny cowbirds, for example, approached the displaying mantis, and some pecked at it. Whether or not it was pecked at, the mantis enhanced its display and struck at the approaching bird with its formidable grasping legs. The cowbird then jumped away and fled to the opposite corner of the cage. Only a few cowbirds managed to kill a mantis, "and this generally happened close to the end of the trial interval (2 hours)." Orioles were more aggressive. Maldonado described encounters between them and mantises:

> No sooner had a bird entered into the cage than it started an attack against the mantid. A real fight between them could be observed. Every time the troupial [oriole] threw a stroke with its beak, the insect answered with a violent enhancement of all the ... [display] components, whereupon the bird suddenly jumped back. An entire series of events succeeded dramatically during a few minutes: displays with noisy stridulation, strikes from the mantid when the foe came near, strong pecks from the bird between attacks and escapes. Finally, the troupial managed to snap at the mantid, throwing it down on the floor, clutched it with one of its feet and brought the insect to the perch, where the prey was eaten. In some trials, however, this series was sharply shortened and the insect was killed at once, because

the attack came so soon that the mantid did not take notice of the foe and did not display . . . on time.

Flash colors and eyespots have been independently adopted—via natural selection—by many different kinds of insects, among them some grasshoppers, mantises, planthoppers, beetles, and moths and butterflies in both the caterpillar and the adult stage. If natural selection favors these warnings, as it seems to, they must benefit these insects in some way—most likely, as we just saw, by helping to ward off predators. This is a compelling hypothesis, which science demands be tested experimentally or by systematic observations. Such tests are difficult with living creatures in nature and even in the laboratory. Nevertheless, there have been some convincing demonstrations that flash colors and eyespots do tend to deter predators.

Observational and experimental evidence supports the hypothesis that the colorful hind wings of underwing moths can startle attacking birds. Theodore Sargent observed captive blue jays attacking underwing moths in an aviary, noting the ways in which the jays seized the moths and the nature of the resulting injury to the wings of moths that escaped from the jays. The characteristics of the injuries varied with the circumstances. For example, moths seized in flight—usually by only one wing—were likely to have a piece torn from just one of the hind or front wings. Birds caught resting moths by their overlapping front and hind wings, tearing corresponding pieces from both wings. The wings of some moths bore no injury except for "crisp" triangular imprints of a bird's beak. This happened, Sargent saw, when a jay momentarily relaxed its grip on a resting moth's wing because it was startled by the sudden appearance of a conspicuous hind wing. Sargent then found similar injuries on free-flying underwings that he caught in the field. The wings of eighteen of seventy-three moths (25 percent) from the field (as against eight of twenty-nine, 28 percent, from the aviary) bore only crisp imprints of a beak, which indicated that the wild moths had likewise escaped from a bird when they flashed their colorful hind wings.

The experiments of Frank Vaughan, one of Sargent's graduate students, showed that the sight of unfamiliar colors caused captive blue jays to hesitate before extracting a mealworm from an experimental feeding device. The mealworm was hidden under a disk in a hole covered by a brown lid. Several jays were trained to push the lid aside and to lift up the disk, which was also brown, to get the mealworm. When later offered mealworms covered by disks of various bright colors, jays hesitated significantly longer before lifting the disk than did jays that were exposed to a disk with a familiar color.

Debra Schlenoff tested the startle hypothesis by presenting artificial models of underwings to caged blue jays. The gray front wings of the models were cardboard triangles, and the hind wings were made of thin plastic painted either plain gray or black with red or yellow bands. The jays were trained to pick up the models to obtain a pinyon nut, a delicacy for them, glued to the underside of one of the front wings. When a jay lifted up a gray front wing, the plastic wing beneath it instantly popped out.

Schlenoff summarized her results: "Jays which had been trained on models with grey hindwings exhibited a startle response when they were exposed to *Catocala*-patterned hindwings. In contrast to this, subjects trained on *Catocala* models did not startle to a novel grey hindwing." Jays that reacted most intensely to patterned hind wings dropped the front wing, raised their crests, flew away, sounded an alarm call, and wiped their beaks. "The startle response to *Catocala* patterns lasted over several days until birds habituated to the models. When the jays had habituated to one *Catocala* hindwing pattern, a novel *Catocala* pattern always elicited a startle response. Familiar *Catocala* hindwing patterns which appeared in an anomalous context . . . associated with a different forewing pattern . . . also evoked a startle response from these birds. Novelty, oddity, conspicuousness, and anomaly are considered as possible stimulus characteristics which trigger the startle response."

But A. D. Blest experimented with real insects. The beautiful peacock butterfly (*Inachis io*) of Europe has four large eyespots, one on the upper side of each of its wings. When the butterfly is at rest with its wings pressed together above its back, only their camouflaged undersides are visible. If the butterfly is alarmed, however, it responds with a defensive display that, as Blest described, "consists of a repeated sequence of movements whereby the wings are [spread,] . . . revealing the eyespots." This display is accompanied by a hissing noise produced by rubbing together certain areas of the wings.

In the 1950s, Blest wanted to determine whether or not this butterfly's eyespots really do evoke "escape responses" in birds. But he was not certain that small birds would attack butterflies or moths "with a wingspan greater than the bird's own length." He found that 6.5-inch-long yellow buntings (now known as yellowhammers) eagerly attacked large wild silk moths with wingspans of 5 inches or more. One of my own experiences confirms Blest's observations. For reasons that I don't remember, I released, in daylight in the city of Urbana, Illinois, a large cecropia moth with a wingspan probably approaching 6 inches. I do remember vividly that I was amazed when a house sparrow, probably less than 6 inches long, seemingly came out of nowhere to snap at the huge flying moth, which, however, managed to escape.

Blest compared the responses of yellow buntings to two groups of butterflies, one with their eyespots intact and another from which the eyespots had been removed by rubbing away patches of scales. The actions of the birds suggest that eyespots do evoke escape responses. When six buntings were tested 159 times, intact butterflies evoked escape responses 128 times, but those from which the eyespots had been removed evoked only 31 escape responses.

Almost fifty years later, in a comprehensive book, *Avoiding Attack,* Graeme Ruxton and his coauthors commented, "Sadly, we are not able to draw very much from Blest's work. as full details . . . are not provided." But they do grant that Blest seems to have shown that eye-spots evoke startle responses from inexperienced birds. They are right. Yet Blest's experiment leaves us in doubt because it did not include a control. For example, in testing a new drug, one group of volunteers is given the drug and another a sugar pill, a placebo, as a control. The intent is, of course, to find out if the drug is more effective than the placebo. Blest's experiment is flawed because rubbing off patches of scales might have some effect other than erasing the eyespots, an effect that might alter the butterfly's behavior in some unknown way that hampers its attempt to startle the bird. The proper control would have been a group of butterflies from which patches of scales equal in size to the eyespots had been removed from other parts of the wings, leaving the eyespots intact. If the birds were as likely to be startled by the controls as by butterflies from which only the eyespots had been removed, we would have to conclude that the removal of the eyespots had little or no effect on the birds' behavior.

In 2005, Adrian Vallin and several colleagues performed a version of Blest's experiment that included a control group. Wild-caught, insect-eating blue tits (relatives of our North American chickadees) were presented with peacock butterflies whose eyespots were covered with paint and a control group with a similar patch of paint on another part of the wing. The results showed that eyespots were "an effective defense; only 1 out of 33 butterflies with uncovered eyespots was killed, whereas 13 out of 20 butterflies with the eyespots covered were killed. The killed butterflies were eaten, indicating that they are not distasteful." Hence, intimidation by bluffing can be an effective means of defense for edible prey. The Vallin group also showed that the hissing sound produced by rubbing the wings together had no effect and that eyespots and sound together and eyespots alone were equally effective. The researchers compared two groups of butterflies with intact eyespots, one group muted by snipping off a small part of the sound-producing mechanism from each front wing. The others, the control group, had an area of similar size removed from each hind wing.

In yet another experiment, Blest placed a mealworm on a horizontal surface. On either side of this bait was an arrangement that allowed him to abruptly project from below an image on a small screen when a bird—a chaffinch, great tit, yellow bunting, or reed bunting—was about to pick up the mealworm. The sudden appearance of circles or concentric circles, which resemble eyes a bit, startled the birds, but they were much more startled by two realistic patterns that looked remarkably like the eyes of a cross-eyed owl. Images that were not at all eyelike— crosses or pairs of parallel lines—had relatively little effect. But Martin Stevens and his coauthors argued that eyespots may be effective not because they look like the eye of a predator but simply because they are conspicuous. There remains, however, the question of why, in nature, there are so many conspicuous circular or oval markings that look like eyes rather than some other conspicuous possibly less startling pattern that does not look like an eye.

While large eyespots probably conjure up a vision of a large and frightening predator to a bird, small eyespots may well suggest a little, inoffensive creature, such as an insect, that might make a meal. Thus, a small eyespot could serve as a target. When a bird pecks at an insect, it aims for the head end, not the tail end, so the insect is less likely to escape. A good way to tell an insect's head from its tail is to look for the eyes. Obviously, a strategically placed eyespot might deflect a peck away from the head to a less vulnerable part of the body, perhaps to a wing or to the tail end. Both Wolfgang Wickler and Denis Owen noted that *Ancyra annamensis,* a planthopper (family Fulgoridae) of southeast Asia with a sucking beak, does an extraordinarily convincing job of making its hind end look like its head end. "The true head can scarcely be seen and is pressed close to the surface on which the bug sits. The conspicuous antennae, the black eyes and the black beak [of the false head] are really appendages of the wing tip," Wickler explained. A bird intending to intercept the fleeing insect by grabbing its head end is likely to aim for the false head and will come up empty-handed when the insect rushes off in the other direction. "The planthopper," said Wickler, "appears to jump backwards when disturbed."

A series of experiments by A. D. Blest indicated that even false eyespots can deflect the aim of attacking birds. "Four hand-reared, inexperienced Yellow Buntings were isolated in small indoor cages; they were accustomed to catching and eating mealworms but had had no experience with eyespot patterns. After becoming habituated to the cages, each bird was presented with a recently killed mealworm that was unaltered or had been painted, some with a 'false head,' a few segments of the hind end painted white that had a small black spot centered in the white area." In a total of 529 trials with unpainted mealworms, the birds attacked the mealworms as expected, with 60 percent pecking at the head end and only 40 percent at the hind end. In another series of trials, with recently killed mealworms painted as described above, the result was quite different. In a total of 430 trials, many of the birds were apparently deceived by the eyespot. More than 65 percent pecked at the false head on the hind end.

On a sunny June day, Jim Sternburg and I watched another deceptive insect whose hind end looks like its head end. We were at Sand Ridge State Forest in west central Illinois, a marvelous sand area and one of the few places in the state where I have seen a cactus, the prickly pear (*Opuntia*), growing. Jim was collecting butterflies, and I was collecting hoverflies, many of which look like bees or wasps and which we will meet again in chapter 10. The deceptive insect was a small butterfly, the banded hairstreak, which was sipping nectar from the white blossoms of a small shrub known as New Jersey tea because tea was made from its leaves during the American Revolution.

This insect is one of a group (family Lycaenidae) of many species, including about one hundred in the United States. Many have one or two long, very thin tails projecting from the rounded rear end of the hind wing. The upper sides of their wings are generally plain brown and only rarely have spots of color. The undersides are usually much lighter, marked with thin hairlike lines, for which these butterflies are named, and usually one or two eyespots at the base of the tail. As Jim and I watched, the hairstreak held its wings together over its back and fascinated us by waving its tails so that they looked like antennae by gently rubbing its hind wings together. While this was going on, the real antennae did not move at all. Jim told me that the many species of hairstreaks he had watched all did the same thing. The false antennae and the eyespots create, as Robert Robbins described, "an impression of a head at the posterior end of the butterfly that diverts predator attacks towards the less vulnerable end of the insect." Lizards, for example, "preferentially attack the false head of [these] butterflies, frequently getting only a mouthful of hindwing while the butterfly escapes almost unharmed." Wickler, succinctly describing Eberhard Curio's observation of a hairstreak (*Thecla togarna*) in Ecuador, said that it "performs a startling feat at the moment of landing: The butterfly turns very quickly so that the dummy head comes to point in the previous direction of flight." Thus this hairstreak misdirects predators both with its false head and by swiftly transposing head and tail ends on landing.

Robert Robbins's systematic examination of 1,400 hairstreaks of 200 different species caught in the wild in Colombia and Panama by him provides convincing evidence that false heads generally cause birds to grab for the wrong end of the butterfly. He began by ranking various species of hairstreaks according to their "predicted deceptiveness" on the basis of the presence or absence of four misleading characteristics: a pattern of lines that cross the wings and converge on the eyespot, as if they were actual wing veins; the headlike shape of the lower angle of the hind wing; the contrast of the false head's color with that of the rest of the wing; and finally, the presence of tails. He then placed species with all of these characteristics in rank I, those with three in rank II, those with two in rank III, and those with one or none in rank IV. As a measure of deflected attacks, Robbins counted the number of wild-caught specimens in each rank that had the imprint of a bird's beak on the hind wings or were missing overlapping pieces from the two hind wings, the latter being evidence that a perched butterfly had torn away from a bird's firm grasp. Specimens of rank I, those with the most-deceptive false heads, were the most likely to have been grabbed by the hind wings, almost 23 percent of 66 individuals. But in the lower ranks, damage to the hind wings steadily decreased, appearing in 12 percent of 293 individuals in rank II; about 5 percent of 554 in rank III; and less than 4 percent of 111 in rank IV, those with the least deceptive false heads.

# SEVEN

# Safety in Numbers

Insects and other animals in groups are less likely to be taken by predators than are lone individuals. Some insects, including most insect eaters—mantises, aphid lions, dragonflies—are solitary. (Except for some ants, I can think of no predaceous insects that hunt in groups.) A lone mantis, for example, will probably escape the very real threat of cannibalism. There is no doubt that mantises are inclined to eat one another; after all, the females sometimes devour their mates. But many insects that are not predaceous—and therefore less likely to be cannibalistic—such as aphids, tent caterpillars, and cockroaches, are gregarious. Group living confers important advantages. Two or more pairs of eyes or ears are more likely than only one pair to sense an approaching predator. A cooperating group of chemically or otherwise armed individuals can mount a more effective defense than a lone individual. Furthermore, stinging or otherwise noxious insects that fend off predators with their bright, conspicuous warning colors have a greater visual impact on a predator if they are in a tight group than if they are alone.

Even if its companions do not act as lookouts or otherwise aid in defense, a member of a group is less likely to be chosen by a predator just because it is only one of many choices. This is commonly known as the dilution effect. A hypothetical scenario makes this point. A lone

Figure 7. A yellow-billed cuckoo eyes a potential meal, tent caterpillars basking in the sun. If the bird becomes threatening, the caterpillars may frighten it away with a cooperative defense.

individual, perhaps a large caterpillar, encountered by a predator too small to eat more than one of these caterpillars will almost certainly be captured and eaten. However, if this caterpillar had been in a group with nine others, all things being equal, there would have been only one chance in ten that it would have become the predator's victim.

Grouping together can benefit insects and other animals not only by reducing their risk of being killed by a predator but also by bringing the sexes together, as noted by Kevina Vulinec. For example, male mosquitoes gather in large, closely packed flying swarms that hover in place, although they may occasionally shift off center by just a few inches. If you sweep a net through a swarm, you will usually catch only males, easily identified by their broad, featherlike antennae, designed to recognize the female's distinctive flight tone. Swarms form because many individual males orient to and hover over the same conspicuous marker, such as a rock at the edge of a pond or a wet spot in a dirt road. Females are attracted to these swarms, but sex-starved males immediately snatch and carry them off to a hiding place in the nearby foliage.

Closely packed aggregations also decrease the rate of water loss. For example, Seigi Tanaka and his colleagues showed that isolated fungus beetles lost 0.48 percent of their weight (essentially equivalent to water loss) per hour, while beetles in groups of 250 lost only 0.22 percent of their weight per hour.

Grouping together enables tiny, newly hatched jack pine sawfly larvae to get at the soft inner tissues of a pine needle if only one breaches its tough outer wall. Arthur Ghent found that a larva that penetrated the outer wall—perhaps because it found a weak spot—was soon joined by others, many of which would have starved otherwise. And if a hungry insectivorous bird comes too close to a large group of sawfly larvae sitting shoulder to shoulder on a pine sprig, it is likely to be scared out of its wits. If startled, these caterpillar-like relatives of the bees and wasps suddenly rear up in unison and discharge from their mouths a drop of odorous, sticky pine resin. The resin, according to Thomas Eisner and his coauthors, "is an effective deterrent to birds."

Demonstrating that the dilution effect in itself reduces an individual's risk of being taken by a predator is, as W. A. Foster and J. E. Treherne explained, complicated because, as we have just seen, "it is likely to be masked by other advantages of group living, such as improved efficiency of feeding and reproduction or, more importantly, by improved detection and confusion of the predator." Foster and Treherne observed solitary individuals and "flotillas" of nymphal marine water striders (*Halobates robustus*) on the water's surface at the shores of the Galápagos Islands in the Pacific Ocean. (This is one of the few insects—you've seen their relatives skating on the surface of freshwater ponds—that live in or on the oceans.) These immature individuals have no interest in mating and apparently do not feed in groups. Because they could not see approaching small predaceous fish that darted to the surface from below, "group size could not . . . improve predator detection or enhance avoidance behavior, such as initiated by the [visible] approach of predators from the air." But Foster and Treherne found that in the absence of "masking factors," it can be seen that group living in itself does reduce an individual's risk of being snatched by a predator. They discovered that "for example, an individual in a group of 15–17 is about 16 times less likely to be attacked than is a solitary individual."

Just a year after Foster and Treherne published their results, Bernard Sweeney and Robin Vannote published an even more convincing demonstration of the dilution effect, which could not possibly have been complicated by masking factors. In South Carolina, small numbers of adult mayflies of a certain species emerge more or less synchronously from a small river just before sunrise, leaving their molted nymphal skins behind. Females mate almost immediately, lay their eggs in the water, and die within half an hour, but males can live for as long as an hour. Most mayflies (order Ephemeroptera, Greek for "short-lived winged ones") survive as adults for a day or two, but a few others live for just an hour or so. In a net stretched across the river downstream from the spawning site, Sweeney and Vannote caught the floating bodies of dead adults and the nymphal skins they had shed when they emerged at the surface. Subtracting the number of adult bodies, which practically always fell onto the river, from the number of nymphal skins, all of which had been left behind on the water's surface, yields a close estimate of the missing adults, which had probably been eaten by the bats, swallows, and dragonflies common along the river. On mornings when swarms included 30 or fewer mayflies, 80 to 90 percent of the adults of both sexes were missing, but when there were swarms of 100 to 250, only 20 to 30 percent of the females were missing, and up to 50 percent of the males, which had been exposed for a longer time.

Some groups form because individuals of the same species are attracted to one another, usually by a chemical signal, a pheromone. First-instar (newly hatched) nymphs of the southern green stink bug (*Nezara viridula*), Jeffrey Lockwood and Richard Story reported, come together in response to their airborne pheromone. The tiny nymphs, which hatch from groups of thirty to eighty eggs that are, according to Robert L. Metcalf and Robert A. Metcalf, laid on a leaf of a host plant (often soybean), stay together; if separated by a disturbance, they release the same pheromone, which is odorless to humans, and reassemble. By pooling their resources, they can secrete a more potent chemical defense, the stink for which they are named. Their pheromone is unusual in that it has two separate functions: in low concentrations it promotes aggregation, but in high concentrations it acts as an alarm signal that triggers dispersal, presumably in response to an overwhelming threat.

Many but by no means all of the four thousand known species of cockroaches are gregarious. Among the gregarious ones is a common household pest, the German cockroach, known to some by the euphemism *water bug*. They are the little tan insects that scuttle for cover when the kitchen lights are turned on at night. "When a piece of filter paper which had been used for a shelter in rearing a group of [German] cockroaches for several days was put into a glass pot in which nymphs had been released," observed Shoziro Ishii, "the nymphs tended to aggregate on this filter paper rather than on clean filter paper. A 3-choice experiment to assess the aggregation of nymphs was carried out using [contaminated] filter paper . . . and 2 clean filter papers of the same size." After forty-three minutes, almost all of the cockroaches had settled on the filter paper that had previously been exposed to cockroaches. "When antennae of the nymphs were cut off, no aggregation was observed. These findings indicate that aggregation was induced by a response to chemical stimuli." Ishii eventually located the stimulus, a pheromone, in the cockroaches' feces.

Groups can form even if the individuals of a species are not in some way attracted to one another. Individuals can also cluster because they develop synchronously, as do mayflies and periodical cicadas, or because, independently of one another, they gravitate to the same resource—perhaps to a food plant, like nymphal locusts, or to a patch of soil particularly suitable for digging nursery burrows, like cicada killer wasps. As we saw, swarms of male mosquitoes come together because many individual males hover over the same marker. If individuals benefit from belonging to such impromptu groups, natural selection is likely to perpetuate their formation, not by favoring individuals attracted to one another but by tightening up the response to a synchronizing stimulus, because predators tend to eliminate stragglers that respond too soon or too late.

A few insects, such as the mayflies of the Mississippi River drainage, periodical cicadas of the eastern half of the United States, and the hugely destructive migratory locusts of North Africa and the Near East, occur in swarms of millions or even billions. So numerous are they that they more than satiate their predators' appetites. Many if not most survive because there are more of them than the combined predators of the area can possibly eat. Thomas Moore, an authority on cicadas, told me that he has watched birds binging on periodical cicadas and once saw a starling so bloated with cicadas that it could not fly. There are, however, exceptional situations. C. Ashall and Peggy Ellis reported that a small group of only a few thousand locust nymphs were totally wiped out in just one week by flocks of starlings, weaver birds, and a few storks.

Swarms of locusts, which are in fact grasshoppers, come and go. They may occur for several years and then disappear for years or even decades. Where were they in the years when no swarms form?

Although locusts have been infamous since biblical times, this question was not answered until 1921. In that year Boris Uvarov showed that what had been considered to be two different species of locusts are actually two phases of one interbreeding species that are very dissimilar in appearance and behavior: a sedentary, solitary phase and a gregarious, swarm-forming, migratory phase.

Locusts switch from the solitary to the gregarious phase when, in a drought year, they crowd together, virtually shoulder to shoulder, on the few surviving patches of vegetation. But what is it about being crowded that stimulates them to make the switch? With a clever experiment, whose results she published in 1959, Ellis showed that the required stimulus is nothing more than being jostled by their companions. She found that even isolated individuals assume the appearance and behavior of the gregarious phase when they are frequently touched by wires dangling from a rotating disk above them. (Recently, a group of researchers at Oxford University did a similar experiment that produced the same results as Ellis's, but they did not mention her publication.) Years after Ellis's famous experiment, Sylvia Gillett discovered that gregarious individuals revert to the solitary phase if they are isolated, separated from their jostling companions. The jostling not only initiates the switch but also, as entomologists put it, serves as an *arrestant* that keeps the members of the swarm together. Almost all, if not all, gregarious animals respond to some arrestant stimulus that keeps the group from dispersing. For example, stink bugs, according to Lockwood and Story, are kept together by the sight, touch, and smell of one another.

In his masterly "Geometry for the Selfish Herd," William Hamilton recognized that even members of defenseless "pacifist" groups gain an extra measure of safety from predators by hiding among their fellows, getting lost in the midst of the crowd, a behavior not unknown among people. (When I was in basic training in the army, Hamilton was a child, and the evolutionary concept of the selfish herd had not yet been formulated. Nevertheless, many of us trainees realized that when our sergeant was choosing victims for such onerous extra duties as KP [kitchen police]—washing dishes, mucking out grease traps—he was less likely to notice us if we were in the middle of the group rather than at its edge.) Insects and other animals confronted by a hungry predator have much more to contend with than a few hours of doing messy work. As is to be expected, evolution has programmed many of them to reduce their risk not only by being gregarious but also by behaving selfishly, competing with their companions for a favorable position in the thick of the crowd.

Because predators tend to attack stragglers at the periphery of a group, Hamilton noted, "most of the herds and flocks with which one is familiar show a visible closing in of the aggregation in the presence of their common predators." For example, "almost any sudden stimulus causes schooling fish to cluster more tightly . . . , and fish have been described as packing, in the presence of predators, into balls so tight that they cannot swim. . . . Similar observations have been recorded for locusts . . . , for gregarious caterpillars . . . and, as various entomologists have told me, for aphids."

Hamilton cited published observations of predators choosing to attack "isolated and marginal individuals" at the outskirts of a group, as do the sparrow hawks of Europe when they encounter a flock of starlings. I once watched a flying flock of starlings in Illinois react to a harrier coursing back and forth close to the ground as it searched for mice in a harvested cornfield. (This hawk is not likely to pursue a flying bird, but the starlings seemed not to know that.) When the hawk was in sight, the starlings, which had been in a loose formation, suddenly and simultaneously came together to tighten their formation and flew almost wing tip to wing tip. The flock looked like a huge amoeba retracting its pseudopodia (leglike extensions). I don't remember starlings shoving their flockmates aside to get into the midst of the flock. But there may well have been some subtle positioning that I didn't notice.

John Hudleston reported some interesting observations of how birds of different sizes go about attacking a swarm of wingless immature locusts (hoppers). "The small birds (e.g., [Old World] warblers and chats) were frightened of dense groups of marching hoppers and tended to feed on stragglers." Only large birds, ravens and hornbills, jumped "into the middle of a roosting, basting or marching group of hoppers." For four days, Hudleston noted, a band estimated to include about two hundred thousand hoppers was under attack by several hundred ravens and hornbills. "The hoppers appeared to disregard the birds while their numbers were large and they kept marching in the same direction, but when they had been reduced to 50,000 they tended to seek the shelter of bushes and grass tufts when attacked." This striking change in the locusts' behavior following their decline in numbers and the probable widening of the space between individuals suggests that they were reverting to the behavior of the solitary phase. This reversal is probably adaptive. When the locusts can no longer hide in the crowd, they hide in the bushes.

Vultures, storks, ravens, hornbills, and other insect-eating birds may follow flying swarms of adult locusts for a few days, but they are opportunists taking advantage of an infrequent occurrence. Some solitary wasps of the genus *Sphex*, however, are apparently nomads that follow locust swarms much as the desert Bedouin follow their flocks of grazing sheep. The wasps provision the burrows in which their larvae develop with adult migratory desert locusts.

In February 1929, C.B. Williams watched a huge swarm of desert locusts land on the ground, followed within fifteen minutes by a great many of these large, black wasps. "Immediately on arrival the [wasps] began to burrow," he reported,

> and a little later dragged paralyzed locusts along the ground into their burrows, laid an egg on each locust, and then sealed the burrow with soil. This continued all that day till dusk, and started again the following morning. . . . Between 1 and 2 p.m. on the second day the locusts on the ground began to fly, and a rapid flight . . . set in: at 2:15 only three live and one dead *Sphex* were seen . . . where two hours previously there had been thousands. They departed in such a hurry that they left hundreds of open burrows,

many half finished; and paralyzed locusts were lying about in dozens, some just alongside the burrows in which they should have been interred.

There is little doubt in my mind that, as Williams proposed, the wasps had left to follow the departing locust swarm, a plentiful source of food for their future progeny.

Only visual, sound, and volatile chemical stimuli can be sensed from a distance. Visual signals rarely foster the formation of groups. With a truly remarkable exception, they are useless in the dark, of necessity are usually simple and bold, and, as Robert Matthews and Janice Matthews explained, "carry less information as distance increases." Insects more often assemble by means of sound signals. Notable among them are the periodical cicadas, which make a deafening din, as Phyllis Cooper and I discovered at Kickapoo State Park during a synchronous emergence of millions of them in east central Illinois. In response to one another's songs, mating choruses of hundreds of males gather in trees during the day and sing incessantly. The louder the chorus, the more females it will attract.

In *Out of Africa* Isak Dinesen wrote of fireflies in the highlands of Kenya: "For some reason they keep within a certain height, four or five feet, above the ground. It is impossible then not to imagine that a whole crowd of children of six or seven years are running through the dark forest carrying candles . . . joyously jumping up and down, and gamboling as they run." But it is in southeast Asia, as John and Elizabeth Buck related, that we can see tens of thousands of fireflies massed in a single tree flashing in perfect synchrony in the night. The males flash at intervals of from about a half second to as much as three seconds, depending on the species, attracting both males and females to the tree. During the brief dark intervals, females emit dim flashes, probably to inform the males that they are ready and willing to mate. After mating, probably with several males (why put all your offspring in one genetic basket?), the females leave to distribute their eggs. Along rivers, such as the Mekong, firefly trees can be seen for miles, and because the same trees are reoccupied year after year, boaters use them as navigation aids at night.

In Costa Rica, Charles Hogue observed that a large tightly packed group of caterpillars of the genus *Hylesia* (family Saturniidae) clinging to a tree trunk responded to the sound of his voice with an intimidating display. "Each of the [caterpillars] responded . . . at the same instant and in the same manner, a violent jerking of the anterior third of the body, so that the head, thorax, and anterior portion of the abdomen were arched upward or sideward." He continued:

> I tested this reaction numerous times and ways by altering the pitch and loudness of my voice and determined to my satisfaction that the action was due to sound and not to air movement. The larvae reacted only to very sharp and relatively high pitched sounds of high intensity. Normal conversation did not cause any movement. I further tested the response by playing music (Strauss waltzes) from a tape recorder in the immediate vicinity of the mass (within 1 meter) and noted that the larvae responded in the same manner to loud, sharp portions of the music.

Hogue noticed that the caterpillars' display tended to discourage parasitic wasps and flies that were approaching to lay their eggs in or on the caterpillars, and that "the high pitched whining of the wings of the approaching or hovering parasite seemed to be of the correct quality and intensity to elicit the jerking response."

A few years after Hogue published his observations, Judith Myers and James Smith observed western tent caterpillars (*Malacosoma pluviale*) in British Columbia using a similar behavior to ward off parasites. On sunny days, large groups of tent caterpillars bask on the outer wall of their silken shelter (see figure 7), where parasitic flies (family Tachinidae) try to glue eggs to their heads. (The maggots that hatch from these eggs burrow into the caterpillar's body and ultimately kill it.) The caterpillars respond to these attacking parasites by violently "flicking their heads" in unison, a behavior elicited by human coughs and the flight sounds of passing bumblebees or attacking parasitic flies. Myers and Smith found that "a tape recording of a flying [parasitic fly] caused head flicking by the caterpillars." Their observations suggest that head flicking is an effective defense. Only 16 percent of the individuals in groups of head-flicking caterpillars on the tent were found to be parasitized, although 52 percent of solitary individuals, which do not flick their heads, had been parasitized.

Both periodical cicadas and the southeast Asian fireflies benefit from being superabundant. The members of a cicada chorus and hundreds of nearby choruses satiate the appetite of attacking predators, the victims saving many—probably most—of their companions, who will give rise to the next generation. It is probably different with the massing fireflies. They, like other fireflies, are unpalatable—probably sickening—to predators, who might sample one or two and then leave in disgust. Their flashing signals may not only bring the fireflies together, but may also be a warning that sickened predators soon learn to heed.

On a sunny day in the Cauca Valley of Colombia, I came upon a group of a hundred or more bright yellow butterflies—relatives of our North American sulphurs and cabbage whites—clustered on a damp spot on the ground. Such groups, also formed by swallowtail butterflies, are often seen in North America, where they are known as puddle clubs, at least to lepidopterists. They form at the edges of puddles or on damp soil—often near feces or where an animal has urinated—to suck up moisture and probably to obtain sodium. When I disturbed the puddling butterflies in the Cauca Valley, they all suddenly swirled into the air. "To a predator," Robert Matthews and Janice Matthews wrote, "such concentrations of brightly colored butterflies could represent a potential bonanza; in fact, they usually do not, primarily because of their behavior. Upon disturbance, masses of butterflies will suddenly fly up and around, surrounding the predator with a whirling cloud of butterflies moving in unpredictable and chaotic patterns, then gradually settling back again only as the source of disturbance wanes. A predator finds it much more difficult, of course, to single out a particular individual among the swirling cloud than to pursue an isolated individual flying away from the group."

With the exception of only one small group, all wasps—both solitary and social—feed their offspring insects and spiders. (The members of the wasp family Masaridae feed them pollen and nectar, as do all of the bees.) Some solitary wasps nest in plant cavities or build mud nests fixed to tree trunks or other surfaces, but others nest in tunnels they excavate in the soil. They stock their burrows with prey they have stung and paralyzed, lay one egg on the prey, and fill the tunnel with soil. Some camouflage their nest sites with bits of twigs and grass, and some—a diverse lot belonging to various genera and families—dig from one to five false tunnels near the closure of each of their nests, most likely, as persuasively argued by Howard Evans, to confuse parasites and predators eager to raid the nests. As Evans said, this behavior probably evolved as an elaboration of the wasps' digging shallow quarries from which to obtain backfill to fill their tunnels. You could say that the wasps create a crowd of false tunnels in which the real nest is lost.

Grouping together is not always the way to minimize the risk of being eaten by a predator. In 1967, Niko Tinbergen, one of the fathers of the modern science of ethology, the biological study of animal behavior, coauthored a study "based on the hypothesis that certain predators exert a pressure on individuals even of well-camouflaged prey species to live well spaced out ... at ... distances which greatly exceed the distance from which predators usually detect them directly." In an area in England covered with low vegetation and frequented by wild carrion crows, Tinbergen and his colleagues laid out plots of hens' eggs painted to resemble camouflaged gulls' eggs. The eggs in one group were spaced 20 inches apart and those in the other about 315 inches apart. "Crows wasted more time searching in the 'scattered' than in the 'crowded' plots, the crowded eggs suffering a much higher mortality."

Families of a few dozen little aphids (plant lice) suck sap from the underside of a leaf or other plant part. Soon I will tell you how the congregation of siblings, all sisters, comes into existence, but first let's see how they scatter when threatened by a predator. An aphid, such as a green peach aphid, that spots a predator emits what Robert Matthews and Janice Matthews called an "alarm-dispersant" pheromone from its cornicles, a pair of thin, tubelike structures jutting up from the hind end of the abdomen. Its sisters respond by slowly moving to the outer margins of the leaf, and some actually drop off the leaf and fall to the ground. With the exception of adult aphid lions (lacewings) and adult ladybird beetles, most aphid eaters—larval ladybirds, aphid lions, and flower flies (family Syrphidae)—are small, slow-moving creepers that can sometimes be escaped even by such a short-distance dispersal.

Groups of aphids form because all of them are the offspring of one mother who gives virgin (parthenogenetic) live birth, at the rate of as many as seven nymphs per day, to a generation of female nymphs, none of which will develop wings. Eventually she produces, again parthenogenetically, a generation of female nymphs that will develop wings. Some stay put, but those of many species fly to an alternate host plant: for example, rosy apple aphids migrate from apple trees, their primary host, to narrow-leaved plantains, common lawn weeds. On plantains a few all-female, parthenogenetic, wingless summer generations are followed in the fall by a generation of winged, sexually reproducing females and the only males of the year. They fly back to an apple tree and mate. Eggs that the females lay in crevices in the bark hatch the following spring. All will produce wingless, parthenogenetic females.

The epitome of group living—and group defense—is practiced by the eusocial (truly social) insects: the honeybees, bumblebees, and a few of the other bees; some of the wasps, such as the hornets and yellow jackets; all of the ants; and all of the termites. In *The Insect Societies*, Edward O. Wilson, father of the science of sociobiology, articulated his insight that these societies "best exemplify the full sweep of ascending levels of organization, from molecule to society." They differ from human societies in that almost all of their members—the workers, plus the soldiers in the termites and some ants—are the nonreproducing offspring of usually just one mother, the queen. They are, so to speak, the queen's troops in the struggle to survive and produce another generation. The worker and soldier castes of eusocial bees, wasps, and ants (order Hymenoptera) consist only of females, which—with the exception of some species of ants—are all armed with a venomous sting they use to defend their colony, swarming out en masse to attack intruding insects and hungry marauders such as skunks, bears, and humans. Termite castes consist of both sexes. They do not have stings, but phalanxes of soldiers do battle with their large, powerful mandibles.

# EIGHT

# Defensive Weapons and
# Warning Signals

Sturdy armor plating, like the iron suits of medieval knights, protects many insects and other arthropods and serves as their supporting framework. This exoskeleton, unlike our endoskeleton, is a more or less rigid body wall, which might be thought of, very loosely, as the insect's skin. Other insects, especially larvae such as caterpillars, have a body wall that is soft and flexible except on the head and legs. But the body wall of most insects consists of rigid plates joined by narrow areas of flexible membrane that make movement possible.

If you have had a gastronomic encounter with the delectable blue crabs of our Atlantic coast, you are well aware that the protective exoskeleton of an arthropod, a crustacean rather than an insect in this instance, can be very difficult to breach. I remember vividly the wooden tables in a crab house on the Delaware shore that were covered with newspaper and supplied with a wooden mallet, a paring knife, a nutcracker, and a big roll of paper towels. Not long after you are seated, a waiter dumps a basketful of boiled blue crabs on your table. You then wield the mallet—not too lightly—to crack a crab's carapace, the nutcracker to crack its claws, and after that the paring knife to pry out its delicious meat. After cracking many crabs, you wipe your face and hands with many paper towels. I have been known to say, only half-jokingly,

Figure 8. A North American skunk and an unpalatable African butterfly independently evolved conspicuous black-and-white color patterns that warn predators of their noxiousness.

that eating hard-shelled blue crabs is so time- and energy-consuming that a person could starve while taking a crack at gleaning their meat.

The majority of the 350,000 known beetle species (order Coleoptera)—about one-quarter of all currently known animals—are among the most thoroughly armored insects. Even their front wings have become hard and opaque, armor plating that covers the membranous, flightworthy hind wings and sheathes much of the body. (In Greek, *Coleoptera* means "sheath wing.") With its forewings (elytra) snapped shut, a beetle is "like an insectan tank, able to resist attack while penetrating all sorts of semisolid [substances]," Stephen Marshall wrote.

One sunny day in July 1982, at the University of Michigan Biological Station just south of the Straits of Mackinac, I netted a colorful soldier fly (family Stratiomyidae) sipping nectar from a blossom. Grasping the fly with my thumb and index finger through the mesh of my aerial net, I could feel it squirming, when, to my surprise, it drove one of its scutellar spines into the ball of my thumb. (The small, shelflike scutellum projects backward from the upper side of the thorax.) I described the most remarkable aspect of this experience in a brief note in the *Proceedings of the Entomological Society of Washington:* "There was enough pain to make me withdraw my hand involuntarily—about equivalent to the prick of a . . . pin." Most likely, the soldier fly's sharp spines would have a similar painful effect on a bird grasping it in its beak, making the bird relax its grip and allowing the fly to escape.

Other insects—including many but not all members of the soldier fly's family—use spines to defend themselves against birds and other vertebrates. Flies of another family (Diopsidae) have similar scutellar spines. Tree frogs spit out stink bugs (family Pentatomidae) when the sharp spines on their shoulders lodge between the frog's jaws. With the spines on their legs, grasshoppers and sphinx moths (family Sphingidae) of some species administer painful scratches. A species of longhorned beetle (family Cerambycidae) of the New World tropics uses the spines on the ends of its antennae to painfully prick the fingers of anyone who tries to hold it.

Many other insects have other physical defenses. Praying mantises strike at attackers with their toothed, raptorial front legs, which are generally used to capture insects. Earwigs (order Dermaptera) can defend themselves by inflicting a painful pinch with the sharp-tipped "forceps" at the end of their abdomen. The insect-eating robber flies and aquatic giant water bugs (family Belostomatidae) are among the insects that can inflict a painful stab with the long, piercing beaks normally used to suck their prey dry. Some beetles and caterpillars bite with their powerful jaws (mandibles). Although almost all insect pupae are virtually immobile, capable only of squirming, those of some beetles and moths are armed with small, jawlike "gin traps" at the segmental articulations of the abdomen, analogous to our mouse traps and other snap traps. When the pupa squirms, they can pinch the leg of an ant or other predatory insect, but they probably give little or no protection against vertebrates.

Insects are good biochemists. They synthesize many chemicals for many uses. Moths, beetles, and many others make glue to stick their eggs to plants; mosquitoes and other blood suckers make anticoagulants to facilitate their feeding; insects of all sorts synthesize a great variety of pheromones that serve a great many purposes, such as attracting mates, marking trails, or broadcasting an alarm. And, as Thomas Eisner and his coauthors made clear in their fascinating and beautifully illustrated book *Secret Weapons*, the defenses of many insects and other arthropods are chemical.

"Chemical warfare," Justin Schmidt explained in his article on Hymenopteran venoms, "is developed to the extreme in the arthropods. There are acids, aldehydes, ketones, quinones, terpenes, alkaloids, plus an almost innumerable diversity [of] other compounds available for organisms to spray onto, daub onto, [inject into,] or vaporize near potential predators." Some insects and other arthropods, as Schmidt pointed out, use "more passive means of chemically defending themselves." They synthesize—or sequester from plants—toxic but usually not lethal compounds that they store in their bodies and that make predators sick when eaten.

Most insects and other animals with effective defenses are conspicuously warningly colored (I discuss this below). Among the few exceptions is the unicorn caterpillar (*Schizura unicornis*) of the north-eastern United States and some other members of its family, the Noto-dontidae. The unicorn caterpillar is exceptionally well camouflaged. As Eisner and his coauthors described, it is "irregularly colored in green and brown, and bizarre in shape. Its dorsal toothlike projections give it a jagged profile that enables it to escape detection by blending in with the [serrated] margins of partially eaten leaves." (Furthermore, as David Doussourd has told us, the unicorn caterpillar is also among those—more of which we will meet in the next chapter—that sometimes prune away partially eaten leaves that might attract the attention of insect-eating birds.) If, despite its superlative camouflage, it is threatened by a bird or other predator, the caterpillar responds by spraying its attacker with a highly repellant secretion consisting mainly of formic acid and a small amount of acetic acid. This repellant, Athula Attygalle and her coauthors reported, is ejected to a distance of 8 or more inches from an opening just behind the head. By raising its head and revolving its front end, the caterpillar can aim the discharge at a leg or other part of its body that is being disturbed—in the laboratory by the pinch of a forceps wielded by an entomologist, but in nature by an attacking predator.

The ability to spray defensive chemical secretions, Eisner and his coauthors noted, evolved independently in many kinds of insects and other arthropods: cockroaches, earwigs, termites, walking sticks, true bugs, beetles, caterpillars, ants, spiders, scorpions, vinegaroons (whip scorpions), centipedes, millipedes, and others.

The vinegaroons, found in the southern United States and the tropics, are named for the odor of their defensive spray, which consists mainly of acetic acid, the chief component of vinegar. They are nonstinging relatives of the scorpions and spiders. Their spray, squirted from glands that, as described by Eisner and his coauthors, open on a "revolvable gun emplacement" at the end of the abdomen, can be ejected for as far as 18 inches and precisely aimed at almost any part of the vinegaroon's body except its underside. In an earlier publication, Eisner and other coauthors reported that the spray repels several insect-eating animals:

wind scorpions, ants, blue jays, Steller's jays, anole lizards, armadillos, and grasshopper mice. The mice, obviously irritated by the spray, tried to wipe it off by pushing their muzzle through sand and by rubbing it with their paws. An armadillo sprayed on its snout responded by "backing up suddenly and jerkily, rubbing its snout against the floor and puffing audibly." Most vinegaroons are not warningly colored. The tailless ones (family Tarantulidae) have black bodies with yellowish markings on the abdomen—but they are not nearly as colorful as many other well-armed insects. Because they hide during the day and are active only at night, a visual warning may seldom be useful to a vinegaroon.

Many of the walking sticks are camouflaged, as is the common twig-like species of the northeastern United States and southeastern Canada. But a species found in the southern United States and the New World tropics, the two-striped walking stick (*Anisomorpha buprestoides*), gives warning of its potent chemical defense with two hard-to-miss bright red stripes that extend the entire length of the upper side of its black body. This walking stick, Eisner and his coauthors M. Eisner and M. Siegler noted, "is the source of one of the most noxious defensive secretions produced by an insect." Writing alone in an earlier publication, Eisner said this secretion causes the eyes to tear up and that "its vapors are painfully irritating when inhaled." The openings of the glands that secrete the potent spray are on the thorax just behind the head and can be aimed in almost any direction. When Eisner used instruments to stimulate various parts of the insect's body, he found that "marksmanship is precise: the spray invariably drenches the particular instrument used for stimulation." Unlike most insects, this walking stick "often discharges its spray preemptively, before it is actually attacked, as by a bird. . . . But it holds its fire until the bird is within a radius of about twenty centimeters [eight inches] or less, well within the range of its spray." In cages the spray effectively deterred ants, beetles, a blue jay, and a white-footed mouse. But a mouse-opossum, although obviously distressed by the spray, persisted until the walking stick was out of ammunition, and then ate it.

When disturbed, the jet-black darkling beetles of the genus *Eleodes* (family Tenebrionidae) assume an unusual posture: supported by their spread legs, they virtually stand on their heads, with the tip of the abdomen held high in the air. This is a defensive posture from which the beetles can spray an attacker, often a bird or a small rodent, in the face with a repellant from openings at the tip of the abdomen. Eisner, Eisner, and Siegler pointed out that these beetles "typically remain motionless in that stance, ready to use their defensive glands should the disturbance persist." This desert dweller's black color is a warning of its chemical defense. It and its predators are active mainly at dawn and dusk, "and there is no better way to be noticeable on sand in a crepuscular setting than by being black." Its stance is also a warning, as witnessed by the fact that a number of defenseless beetles try to fool predators into believing that they have a chemical defense by mimicking both the darkling beetle's color *and* its defensive stance. The darkling beetle's spray is an effective deterrent except against "the grasshopper mouse . . . , which has the remarkable, probably learned habit of holding the beetle upright . . . and forcing its rear into the sand, thereby causing the secretion to be discharged ineffectually into the soil." The mouse eats the whole beetle from the head down, except for its hard wing covers and the tip of the abdomen, which contains the repellant-secreting glands.

In John Steinbeck's *Cannery Row,* Doc, a marine biologist with a philosophical bent, and Hazel, a friendly tramp, come upon a group of darkling beetles. Hazel asks Doc, "What have they got their asses up in the air for?"

"I don't know why," [Doc] said. "I looked them up recently—they're very common animals and one of the commonest things they do is put their tails up in the air. And in all the books there isn't one mention of the fact that they put their tails up in the air or why."

Hazel turned one of the stink bugs over with the toe of his wet tennis shoe and the shining black beetle strove madly with floundering legs to get upright again. "Well, why do *you* think they do it?"

"I think they're praying," said Doc.

"What!" Hazel was shocked.

"The remarkable thing," said Doc, "isn't that they put their tails up in the air—the really incredibly remarkable thing is that we find it remarkable. We can only use ourselves as yardsticks. If we did something as inexplicable and strange we'd probably be praying—so maybe they're praying."

A toad slings out its long, sticky tongue to snap up a half-inch-long beetle. The toad doesn't know it yet, but that was a big mistake! When the toad is about to pull the beetle into its open mouth, it suddenly gags, releases its prey, gapes widely, and rubs its tongue against the ground. The beetle, a North American species of bombardier beetle (genus *Brachinus,* family Carabidae), had blasted a noxious spray at the temperature of boiling water (212°F) from the tip of its abdomen directly into the toad's mouth. After a few unpleasant encounters with these beetles, the toad will learn to leave them alone, recognizing their warning coloration—a reddish thorax and iridescent blue wing covers—and associating it with their noxious defense. People are also sensitive to a bombardier beetle's explosive charge. Harold Bastin quoted an entomologist, J. O. Westwood, who reported in 1839 that when seized, individuals of a large South American species of bombardier beetle "immediately began to play off their artillery, burning and staining the flesh to such a degree that only a few specimens could be captured with the naked hand." In North America, Stephen Marshall noted, bombardier beetles "are often common under shoreline debris especially near bodies of water frequented by whirligig beetles (Gyrinidae). The larvae of bombardier beetles are parasitoids of whirligig pupae, which are found under domelike mud shelters along shorelines."

How on earth do these beetles manage to produce such a repugnant, boiling hot, explosive discharge? They accomplish this with one of the most elaborate defensive weapons in the insect world. The bombardier beetle's abdomen, as described by Eisner, contains glands that store hydrogen peroxide and hydroquinones in two separate compartments.

When the beetle is ready to release a blast, these chemicals are forced into another chamber, where enzymes cause them to react explosively. This reaction results in the production of heat and highly irritating quinones and blasts them out through a duct at the tip of the abdomen. "Through the rotation of the abdominal tip," Daniel Aneshansley and his coauthors noted, "they can eject the spray in virtually any direction, and they always aim it toward the foe."

Some animals have stingers or fangs, essentially hypodermic needles, that they use to inject venom into other animals. The most familiar are snakes and insects such as ants, wasps, and bees, but there are others, among them spiders, scorpions, centipedes, and the marine snails known as cone shells. Most sting primarily to subdue prey and only secondarily as a defense against predators. In "Hymenopteran Venoms: Striving toward the Ultimate Defense against Vertebrates," Justin Schmidt wrote, "Venoms are not a defensive panacea for stinging Hymenoptera. However . . . [they] are the next best thing. [They] have given their possessors the ability to explore and exploit many new habitats that otherwise would be closed or risky to them because of potential vertebrate predators." The order Hymenoptera includes the only stinging insects, the ants, all of which are social, and the wasps and bees, a few of which are social. Only female Hymenopterans have stingers, which are highly modified ovipositors. Their venoms, Schmidt pointed out, "owe their defensive utility" to the stinger, which pierces the vertebrate skin and delivers them "into sensitive tissues, where their activities of pain, tissue damage, and lethality can occur."

The one and only queen in a colony of common honeybees (*Apis mellifera*) does almost nothing but lay eggs. The workers, her daughters, are all sterile. They do all the chores of the colony: among them, air-conditioning the hive, rearing the young, and foraging for pollen and nectar, the colony's only foods. They use the stinger only to defend the colony against invading insects—and marauders such as bears, humans, and other vertebrates, which exploit honeybee colonies as bounteous sources of food. Queens sting only to kill competing queens in internecine battles in the hive. (In spring, the queen and a swarm of workers leave the hive to found a new colony elsewhere. A new queen, one of several whose development had been initiated weeks before by the workers, succeeds the absconded queen after stinging to death the other new queens.)

Guard bees at the entrance to the hive, Charles Michener explained, are the first line of defense. When they sense danger, they raise the abdomen and expose the stinger, releasing into the air an odorous alarm pheromone from its base, and vibrate their wings to disperse it. In response to the alarm pheromone, other workers, hundreds or even thousands, rush forth to sting the marauder. The entire stinging apparatus, anchored by the barbed stinger, is usually ripped from the bee's body as she flies off, but it continues to pump venom into the enemy and to release the alarm pheromone. The attack rapidly escalates as more and more workers release the alarm pheromone. The marauder may be driven away or even killed by multiple stings, especially if it is a human rather than a bear protected by a thick coat of hair.

In *Robbing the Bees,* Holley Bishop told us how honeybees have been used as weapons of warfare, from the time of the ancient Greeks and Romans until the twentieth century. "Before sophisticated box hives were invented, bees kept in twig, straw, or clay vessels were adapted as weapons ... [that] could be hurled through the air or dropped on enemies." When attacking medieval castles, the besiegers, using catapults, hurled hives of bees over the walls. "By the time they reached their target, the ... bees were outraged, ready to explode in a fury of stings."

The African race of honeybees, sometimes called killer bees because of their exceptionally belligerent defense against vertebrates, are more likely to launch deadly mass attacks than other races of *Apis mellifera.* They were accidentally released in Brazil in 1957 by apiarists who hoped to hybridize the more productive Africans with the Italian race they kept, in order to increase the yield of honey. Since then the Africans have moved as far north as the southern United States and have interbred with the much less belligerent domesticated Italian race. These Africanized bees have killed hundreds of people in South and Central America but so far, May Berenbaum, who is well acquainted with bees, told me, only about forty in the United States.

When I was an undergraduate at the University of Massachusetts, I had a fantastic summer job. My partner and I traveled all over New England in a truck, the Covered Wagon, presenting nature study programs to children in summer camps. During a nature walk with a small group of boys, one of them stepped on and caved in the underground nest of a colony of yellow jackets (family Vespidae), which often build their paper nests in abandoned animal burrows. Scores of angry wasps immediately came to the defense of their colony. I yelled, "Let's get out of here!" All the boys ran, except for one too terrified to move. Both of us were stung many times as I carried him to safety. Unlike honeybees, yellow jackets and other wasps can sting repeatedly.

The stings are instantly and intensely painful, sufficient to drive off not only people but also other marauding vertebrates. Yellow jackets, like all social insects, have much to protect. A colony of ants or wasps will contain hundreds or even thousands of larvae and pupae, a bountiful meal that may tempt animals as large as bears. A colony of honeybees is even more tempting because in addition to larvae and pupae, it usually contains large stores of honey and pollen. The social wasps (including yellow jackets), as Roger Akre and his colleagues explained, use their stings mainly as defensive weapons. They inject venom laced with substances that cause intense pain in vertebrates and, like honeybee and ant venoms, also contain components that damage vertebrate tissues and in large quantities may even be lethal.

But, as Akre and his colleagues pointed out, most stinging wasps and bees are not social. These solitary wasps use their stings primarily for subduing the prey that they, like all social wasps, provide for their larval offspring. Their venom is lethal to their prey but in most cases causes only slight and fleeting pain in people, as does the venom of solitary bees. The existence of almost-pain-free venom in these solitary wasps and bees makes perfect evolutionary sense because they seldom or never need to defend their nests against vertebrates. Hunting for the relatively tiny amounts of food in the scattered nests of solitary wasps or bees will be a losing proposition for almost any large vertebrate.

There are a few exceptional solitary wasps that do have very painful stings. For example, wingless female wasps known as velvet ants (family Mutillidae) because of their antlike shape and dense "pelage"—a covering of short, often red, erect hairs—are a temptation for birds as they scurry about conspicuously on the ground looking for the entrances to the burrows of other wasps or bees, into which they will drop the eggs from which parasitic larvae will hatch and devour the wasp or bee larvae. Many velvet ants are warningly colored, including a large one known as the cow killer (*Dasymutilla occidentalis*) because of its excruciatingly painful sting. Its warning is unmistakable: the upper side is bright red and bedecked with red hairs.

"Until recently," Bert Hölldobler and Edward O. Wilson wrote, "the search for the ancestry of the ants always ended in frustration." This changed when a perfectly preserved fossil of the most anatomically primitive and oldest known ant was found embedded in eighty-million-year-old sequoia amber from New Jersey. Like many fossils embedded in tree sap, this long-dead ant looks as if it is alive and could walk away if it were freed from its prison (see Hölldobler and Wilson's plate 1). It is a no-longer-missing link between the wasps and the ants, a member of a now extinct, previously unknown group from which all of the ants (family Formicidae) sprang.

A long extruded stinger is perfectly obvious in this fossil. Most living ants, the more primitive species, also have functional stingers, which, generally speaking, they use as a weapon both for killing and paralyzing prey and for detering marauders that threaten their nest. Most of the less primitive, more recently evolved species, however, have nonfunctional vestigial stingers and therefore must defend themselves with other forms of chemical warfare, such as emitting droplets or sprays of poisonous substances. An extreme example of this latter type of chemical warfare is the "bomb defense" of a Malaysian species (*Camponotus saundersi*). During combat, a suicidal worker, according to Hölldobler and Wilson, "contracts its abdominal wall violently, finally bursting open to release" from a hugely enlarged internal gland a large mass of a poisonous secretion that is likely to hit the marauder.

The most infamous of the stinging ants in the United States are, without doubt, the fire ants (genus *Solenopsis*), named for the sharp, fiery pain of their sting. The red imported fire ant (*Solenopsis invicta*), a native of South America, was first found in the United States in the first half of the twentieth century and has since spread throughout most of the southeastern states. The nests of this ant are hard-crusted earthen mounds up to 3 feet in both diameter and height and may number from twenty-five to one hundred per acre. A mound may house one hundred thousand or more fiercely stinging worker and soldier ants, all females. If their nest is disturbed, hordes of agitated workers and soldiers pour out and aggressively attack intruding humans or other animals. An attacking worker grips the flesh with its powerful mandibles for leverage and then drives in its sting. Berenbaum described the effect of its sting on a person:

> The consequences of fire ant stings are dramatic. An immediate burning sensation at the site of the sting. . . . A swelling soon appears and a blister or vesicle forms at the site. Within a day, the vesicle fills with pus as venom constituents break down cells and tissues. Within a week the pustules are reabsorbed, often leaving behind scar tissue. Systemic reactions can also occur in response to the sting and can involve nausea and vomiting, disorientation and dizziness, asthma, and other allergic responses; in some cases, stings can be fatal.

There is no doubt that fire ant stings are an effective defense against vertebrates. Fire ant venom not only has a powerful effect on humans but often kills small birds and mammals.

There is a fascinating footnote to the story of the red fire ants' arrival in the United States. The various texts I have consulted do not agree on when this ant was first discovered here. One says that the year was 1918; another says "some time prior to 1930"; and another, which comes close, says that this fire ant "was first found at Mobile, Alabama, about 1940." I did not learn the true story until I read Edward O. Wilson's *Naturalist*. Wilson is the foremost expert on ants and one of the founders of the science of sociobiology. He has been awarded two Pulitzer Prizes for his books on ants and sociobiology.

In *Naturalist,* Wilson traces his development as a naturalist and a biological scientist from his boyhood on the Gulf Coast of Alabama to his years as a professor at Harvard University. When he was seven years old he wandered a beach near his home in Mobile looking for shells and scanning the waves for the sight of a shark's fin. In the fall of 1942, when he was thirteen, he "set out to collect and study all the ants in a vacant lot next to" his home. Among his discoveries were some colonies of red fire ants. "The vacant lot discovery was the earliest record of the species in the United States, and I was later to publish it as a datum in a technical article, my first scientific observation," Wilson wrote.

A stinger is not the only way to inject digestive enzymes and/or venom into an attacking predator. Many different kinds of insects with piercing beaks, usually used to extract body fluids from their insect prey, also use them to stab and inject venom into attacking vertebrate insectivores. For example, although some assassin bugs (family Reduviidae) suck blood from vertebrates, most prey on insects. In either case, they use their mouthparts to pierce the bodies of their victims and suck up vertebrate blood or insect blood and tissues that have been, as Vincent Wigglesworth explained, liquefied by enzymes in the saliva that the assassin bug injects into its prey. An African insect-eating species of assassin bug (*Platymeris rhadamanthus*), according to J. S. Edwards, can spray a defensive discharge of saliva up to 12 inches, causing intense pain, especially in the sensitive tissues of the eyes and nose. But like most if not all assassin bugs—including North American species—it can also stab when attacked, piercing the vertebrate attacker with its mouthparts and injecting painful salivary fluid. Although most assassin bugs are an inconspicuous gray or brown, others are warningly colored, often with red and black.

According to M. Deane Bowers, the bodies of some caterpillars of twelve families of moths and one family of butterflies bristle with stiff, venom-loaded breakaway hairs or spines. Upon contact with an attacking predator, these hairs pierce its skin and snap off, releasing their venom. From personal experience I know that the resulting pain resembles that caused by brushing against the leaf of a stinging nettle but is likely to be much more intense and irritating. The caterpillars' venom-filled hairs are said to be urticating, from the Latin word *urticare*, "to sting." (The scientific name of the nettle plant is *Urtica dioica*.)

In chapter 6 we became acquainted with the adult North American io moth, which startles predators by suddenly exposing the large, glaring eyespots on its hind wings. But in the caterpillar stage it is protected by closely spaced, rosette-shaped tufts of exceedingly sharp, pale green, black-tipped urticating spines on its back. You might encounter and brush into an io caterpillar on almost any kind of plant. The late John Bouseman and James Sternburg reported that they feed on "most species of deciduous trees and shrubs, and . . . coarse grasses including corn." Young io caterpillars are conspicuous because they are orange and gather together in groups while clinging to a plant. The older ones, green and solitary, are usually hard to see from a distance, but close up their urticating spines are clearly visible, as is the warningly colored red stripe edged with white on each side of the body. "Anyone who brushes against an io moth larva with bare skin," noted Thomas Eisner, M. Eisner, and M. Siegler, "will experience instantaneous localized pain, followed by itching and inflammatory swelling. . . . Given their effects on humans, there can be no doubt that the spines are protective against natural enemies."

"The irritant hairs with ... the strongest effect on human skin," Walter Linsenmaier pointed out, "are those of the caterpillars of certain flannel moths [family Megalopygidae].... In Brazil the larvae are known as *bizos de fuego* ('fire beasts'), in Paraguay as *iso jagua* ('jaguar worms')." As unnerving as these names are, they fall short of expressing the excruciatingly painful—often debilitating—effect that brushing against certain of these caterpillars can have. The most dangerous (genus *Megalopyge*) in the New World tropics are about 2 inches long and half that wide. With its body completely hidden by a dense pelt of long, swept-back yellow or white hairs, a larva sitting in plain sight on the upper surface of a leaf, as flannel moth larvae often do, looks like a toupee for a tiny elf. The toupee, however, conceals dangerous urticating hairs. I encountered several of these "fire beasts" in Colombia but fortunately never brushed against one, although I did hear about an American ecologist who did. The pain was immediate and so excruciating that he had to be hospitalized for a few days.

Caterpillars of some species do not wastefully discard their urticating hairs with the skin they shed when transforming to the pupal stage. Before shedding they salvage many of these stinging hairs to protect themselves during the pupal stage and the succeeding adult stage. The adults of some species even cover their egg masses with these already thrice-used urticating hairs.

The brown-tail moth (*Euproctis chrysorrhea*), an import from Europe that is often a destructive defoliator of trees in North America, practices this marvelous parsimony. In 1913, Harry Eltringham described how a full-grown brown-tail moth caterpillar incorporates urticating hairs as it spins the silken cocoon that will house it during the pupal stage, weaving them into the inner and outer walls. According to him, "the inner lining of the cocoon is of much looser silk, and though spicules [urticating hairs] are scattered all through it there is a particularly dense mass of these arranged roughly in a belt round the inside of the lining, and placed toward the anterior end, a little beyond the middle." Peeking through a hole he had cut in the hind end of a cocoon, Eltringham observed how an emerging female entangles urticating hairs in a tuft of ordinary, nonurticating hairs at the anal end of her abdomen:

> The anterior end of the cocoon seems to be comparatively thin, and a thrusting movement of the head and thorax soon tore a hole through which the moth emerged and ran to the side of the box. A female ... at first emerged only so far as the anterior part of the thorax. In this position the extremity of her abdomen was just on a level with the band of spicules in the cocoon and she proceeded to carry out the peculiar movements I have already described. The anal tuft could be distinctly seen moving round and round the cocoon and opening and shutting amongst the spicules.

The conspicuous white-winged female brown-tail moth lays her eggs in masses on the undersides of leaves and covers each mass with a protective shield of urticating hairs by pressing down upon it with the tuft of hairs at the hind end of her abdomen.

This caterpillar's urticating hairs cause a severe rash in humans and may even kill, at least in fiction. Years ago I read a mystery story in which a person became chronically ill and ultimately died by inhaling hairs of brown-tail moths that a murderer repeatedly fed into a ventilation system. Unfortunately, I can't recall the author's name or the title of the story. It should be easy to avoid brown-tail moths and their cocoons, as most insect-eating birds do, according to E.B. Ford. The caterpillar is strikingly aposematic (warningly colored), dark brown with a white stripe on each side of the body and bright red tubercles on its back. It is, as Ford described, "a striking object" when it feeds in full view upon the leaves of its host tree.

Some insects have internal glands that can be protruded from the body in response to a threat from a predator. In their usual involuted position, the glands contain the fluids that they secrete. When everted, literally turned inside out like a sock, their inner surface, coated with these deterrent and often volatile fluids, is exposed to the air. Among these insects is the tiny 0.2-inch-long black larva of the willow leaf beetle (*Plagiodera versicolora,* family Chrysomelidae), unintentionally introduced with cargo into North America from Europe in 1911. The larvae are conspicuous as they feed on the upper side of a willow leaf, partially skeletonizing it by eating its upper surface but leaving the veins intact. If gently pinched with forceps, a larva everts all or some of its eighteen defensive glands, arranged in rows of nine on either side of its body. The secretion of these glands is, Eisner, Eisner, and Siegler reported, "potently repellent to arthropod predators," and I presume to birds and other vertebrates.

Caterpillars of the swallowtail butterflies (family Papilionidae) defend themselves with an eversible glandular organ, the osmoterium (from Greek roots that combine to mean "an instrument for disseminating odors"). An alarmed swallowtail caterpillar, Hazel Davies and Carol Butler noted, protrudes the Y-shaped osmoterium, which "is otherwise hidden in a pocket behind the head." When the glands, the two arms of the Y, are everted, "they extend and swell like pushing out the fingers of a glove." The osmoterium is bright red or yellow, depending on the species, and gives off a strong odor that, Eisner and his coauthors, mentioned above, remarked, has been shown to repel "both invertebrate and vertebrate predators."

Other insects likewise produce chemically complex defensive substances but have more "low-tech" ways of discharging them. Some regurgitate them, while others just ooze them from a part of the body. Douglas Whitman and his coauthors listed twenty-two families of six different insect orders that include some species that regurgitate when disturbed. Regurgitated plant matter may contain obnoxious plant substances, and an insect may have glands that add its own defensive compounds to the mix. Among the regurgitators are many caterpillars, including those of the polyphemus moth (*Antheraea polyphemus,* family Saturniidae) of North America. This handsome giant silk moth, which may have a wingspan of as much as 6 inches, is well known to amateur naturalists, who collect its cocoons in winter and take pleasure in the adults that emerge from them in spring.

Among the leaves of the oaks, birches, and other trees on which they feed, the big—up to 3 inches long—green polyphemus caterpillars are well camouflaged and difficult to spot. But if disturbed by a sharp-eyed insect eater, they often respond by first making clicking sounds and then regurgitating a noxious brown fluid. Although this behavior was first observed more than one hundred years ago, its purpose was not understood until 2007, when Sarah Brown and her coworkers published the results of their painstaking and comprehensive research. They found that pinching a polyphemus caterpillar with forceps usually prompts it to make a series of clicks by scraping its mandibles against each other. If the researcher persists in pinching, the caterpillar regurgitates. The clicks are loud close up, but they carry for only a short distance, just far enough to warn a threatening bird of the caterpillar's noxious chemical defense but not far enough to alert more-distant predators. Brown and her colleagues demonstrated that the regurgitant, if smeared on food, inhibits the feeding of both ants and house mice. They also found that ten other caterpillars—some giant silkworms and the others sphinx moth larvae—regurgitate when pinched, but only three of them, one silkworm and two sphinx caterpillars, produce warning clicks before regurgitating.

If disturbed by a predator or a researcher, blister beetles (family Meloidae) ooze from their leg joints blood containing cantharidin, a potent irritant that causes a severe blistering of the skin of humans and other vertebrates. Certain blister beetles, including some North American species, are warningly colored—some a shining metallic blue and others a combination of red with black or dark brown. Most vertebrates, according to Eisner, Eisner, and Siegler, will not eat blister beetles, but, amazingly, a few, including quail and the European hedgehog, do eat them and suffer no ill effects. Only male blister beetles can make cantharidin, which they transfer to females during copulation. This gift may increase the male's own reproductive success, his evolutionary fitness, by enhancing his mate's ability to ward off predators, and because she will protect her eggs—fertilized by his sperm—by transferring to them some of his cantharidin. The infamous Spanish fly, a misleadingly named blister beetle, is said to be an aphrodisiac. But it is dangerous to use: even a very small quantity of cantharidin may be lethally toxic if ingested.

While strolling the boardwalk of the Anhinga Trail in Everglades National Park, I watched a purple gallinule walking slowly on floating water lily pads, nipping up all insects within reach. But the bird ignored an eastern lubber grasshopper (*Romalea guttata*) resting on a leaf, stepping right over it, most likely because its flashy warning colors, yellow and black, reminded the gallinule of a previous unpleasant encounter with the noxious chemical defenses of another lubber.

This large, heavy-bodied insect does not leap away from predators, as do other grasshoppers. "It gives the impression, often conveyed by well-defended animals, that it," as Eisner, Eisner, and Siegler so aptly described, "has nothing to be concerned about." It plods along slowly, does not hide, and if disturbed neither leaps nor, being short-winged, flies away. When intruders approach, lubbers augment their warning display by rising on their legs, lifting their front wings, and partly raising their bright crimson hind wings. This is not a bluff. If pecked at by a bird, the grasshopper raises its hind wings the rest of the way, flails with its spiny hind legs, regurgitates a noxious fluid, and—with a hiss—explosively discharges a malodorous froth from an opening on either side of its thorax.

Murray Blum, now an expert on insect chemical defenses but once my fellow graduate student at the University of Illinois, told me that birds, mammals, and reptiles are obviously frightened by this display and quickly back off. A bird that eats a lubber becomes ill, vomits, and will thereafter refuse to so much as touch one of these insects. But according to Reuben Yosef and Douglas Whitman, some birds can cope with lubbers. Loggerhead shrikes, also known as butcher birds, impale them on a thorn, as they often do with other prey, and don't eat them until a day or two later—after some of the grasshopper's defensive chemicals have degraded. Even so, they don't eat the thorax, which is the seat of the lubber's chemical defense.

Some of the true bugs (order Hemiptera) direct a spray of defensive chemicals at their enemies, as do the vinegaroons and walking sticks we met earlier in this chapter. But as Heinz Remold related, if attacked, the nymphs of certain seed bugs (family Lygaeidae) spray defensive chemicals on their own backs "in order to surround themselves with a protective covering." Remold also noted that other Hemiptera, among them some adult leaf bugs (family Miridae), brush chemical agents on aggressors with their feet, first moistening the foot closest to the point of attack (near the researcher's forceps, say) by wiping it against the opening to the poison gland.

Some kinds of insects contain internal toxins that are usually not lethal and affect a predator only if it eats one or more of them. I know what you are thinking. I will eventually get around to the question of how an insect could possibly benefit from being eaten. But first let's consider how the bodies of some—but by no means all—insects come to contain such toxins. Some synthesize their own; others sequester toxins they obtain from their food plants or even their prey. The scientific literature often says that these toxins make insects unpalatable—distasteful or disagreeable to eat. A bad taste alone, however, is not likely to deter predators for long. After all, humans acquire a taste for anchovies or hot chilis. Natural selection will favor insects that acquire a taste for a nontoxic, nutritious food whose flavor they had at first considered bad. We must add *sickening* to the definition of these toxic but usually nonlethal chemicals.

Douglas Whitman and his coauthors tallied beetles of eight families and moths and butterflies of twenty-one families that "autogenously derive"—synthesize on their own—a variety of toxins. The beautiful aposematically colored, day-flying burnet moths (family Zygaenidae) of Europe secrete hydrogen cyanide (HCN), a deadly poison, and store

it in cavities in their exoskeleton. When the late Lady Miriam Rothschild, a competent entomologist, told the eminent English insect physiologist Sir Vincent Wigglesworth that these moths have this toxin, he,
like many others, did not believe her, declaring that "he would eat his
hat if it was proved true that the burnet moths really contained HCN!"
He was wrong, but I don't know if he ate his hat.

Many aposematic insects—probably thousands—sequester plant
toxins in their bodies. They include members of several orders and
various families, among others, moths, beetles, true bugs, aphids, and
grasshoppers, which among them sequester toxins from quite a few different plant species of various families. Ritsuo Nishida summarized
many of the known examples of just nine moth and eight butterfly families that include species that sequester plant substances, which he says
"are presumed to play decisive roles as defensive agents."

Why do plants contain the toxins that so many insects sequester in
their bodies as weapons against their own predators? Worldwide, there
are well over 260,000 species of seed plants. A few insects feed on other
kinds of plants, such as mosses and ferns. But the species of seed plants
are greatly outnumbered by the species of insects that feed on them,
more than 430,000 different kinds—almost 46 percent of all the known
insects and about 36 percent of the known animals. As is to be expected,
the plants have evolved many different ways—sometimes physical,
such as thorns or hairs, but more often chemical—of defending themselves against this great horde of enemies. As a group, the seed plants
produce a huge variety of toxins, many thousands of them, to deter
insects and other animals that intend to nibble on them. Milkweeds,
for example, contain cardiac glycosides, also called cardenolides, that
are toxic to both mammals and insects. (As their name indicates, some
of these cardenolides, such as digitalis, are used in very low doses to
treat heart problems in humans.) Wild cherry trees contain amygdalin, a cyanogenic glucoside, which gives rise to deadly hydrocyanic
acid when digested by insects or vertebrates. According to L. R. Tehon,
C. Morrill, and R. Graham, fresh cherry leaves do not contain enough

amygdalin to poison browsing cows, but the increased concentration in wilted leaves on lopped-off branches is enough to kill them.

Some insects, such as grasshoppers, armyworm caterpillars, and Mormon crickets, are generalists that feed on many different species of unrelated plants. But the majority of herbivorous insects are host plant specific. They feed on a limited selection of plants, often only a few closely related species that belong to the same family. These host-specific insects have evolved immunity to the chemical defenses of the plants on their menu; they detoxify or store the toxic substances somewhere in their bodies where they can do no harm.

Among these host plant–specific insects is the beautiful and beloved monarch butterfly (*Danaus plexippus*), which in the caterpillar stage feeds on nothing but the leaves of various species of milkweed plants. Not only are monarch caterpillars not poisoned by the milkweeds' cardenolides, but they sequester them in their bodies and pass them on through the pupal stage and into the adult stage. These toxins are seldom lethal to birds or other predators that have eaten a poison-laced insect—for an evolutionarily significant reason that I will soon come to. Monarch butterflies, for instance, contain an emetic dose of cardenolides that is somewhat lower than the lethal dose for a bird. In other words, the chemical will make birds vomit, purging themselves of the poison before it can kill them.

The oleander aphid (*Aphis nerii*) sucks sap from oleander leaves and sequesters cardenolides in its body, according to Miriam Rothschild and her coauthors. Seven-spotted ladybird beetles (*Coccinella septempunctata*) that eat these aphids contain no cardenolides, but according to a subsequent article by Rothschild, the eleven-spotted ladybird (*Coccinella undecimpunctata*), which also eats these aphids but never any part of the oleander plant, does sequester cardenolides. It gets them secondhand, from the bodies of the aphids. In another article, Rothschild and the same coauthors report that a lacewing (*Chrysopa*) also sequesters cardenolides from the scale insects that it eats.

"Fireflies," Thomas Eisner and his coauthors noted, "are chemically protected. Species of the genus *Photinus* . . . contain certain distasteful steroids that protect both sexes from spiders, birds, and probably other predators." These steroids are "also present in a firefly of the genus *Photuris* . . . but in substantial quantity in the female alone and, oddly, only after the mating season was well underway." The reason for this, as James Lloyd found, is that the carnivorous female *Photuris* attracts males of other species of fireflies by imitating the flashes—the mating signals—of females of the victims' own species. They have broken the *Photinus* code. These femmes fatales then devour the unfortunate amorous males. The *Photuris* females get not only a meal but also the distasteful steroids from their victims' bodies, which they sequester.

Although most insects are generally secretive and camouflaged, those that have effective ways of defending themselves against predators are usually, as we have seen, far from secretive. In fact, they flagrantly advertise their presence with conspicuous warning colors. Monarch butterflies, for example, have bright orange wings striped with black. In biological parlance they are said to be aposematic.

Aposematic color patterns almost invariably contrast with the animal's background and are typically black, white, red, yellow, orange, or a combination of two or more of these colors. These patterns are out of the ordinary in nature, aide-mémoire that help predators recognize and steer clear of insects and other potential prey that have in the past made them ill or caused them misery by stinging, spraying, or otherwise inflicting distressful or painful but seldom lethal substances. Not surprisingly, people use these same colors as warnings. Along city streets and highways, stop signs and stoplights are red; signs that warn of tight curves or other dangers ahead are yellow and black; the maritime storm warning flag is a black square on a field of bright red.

Some biologists take *aposematic* to mean nothing more than warningly colored. Insects that are toxic or otherwise defended are, to be sure, almost all brightly, eye-catchingly colored, but, generally speaking, most also have other ways of attracting notice to themselves. Attention-grabbing sounds or movements often accompany warning coloration. When sipping nectar from a blossom, warningly colored wasps make themselves even more conspicuous by rocking from side to side. A periodical cicada produces a loud "shriek" if grasped by the fingers or, presumably, a bird's beak. In addition, as Graeme Ruxton and his coauthors point out, "many chemically defended insects utilize ... odours in combination with visual displays." Tiger moths (family Erebidae) emit warning signals of three different modalities—vision, odor, and sound—to proclaim their toxicity. They are, Stephen Marshall noted, "among the most brightly colored and distasteful of moths." Miriam Rothschild and P. T. Haskell described the aposematic signals of a European garden tiger moth (*great tiger moth* in the United States), *Arctia caja:*

> A touch on the underside delivered from below frequently results in the moth falling to the ground with its legs drawn in and up, in such a manner that the maximum amount of red is exposed, and the wings folded, "shamming death." ... If ... the moth is jarred or touched from above ... there is an immediate release of strong-smelling defensive vapour from the cervical [neck] glands ... and the wings are spread so as to expose the scarlet hind areas. ... The initial emission from the defensive glands is colourless, but the mixing of haemolymph [blood] and air with the secretion turns it into a frothy mass of yellow bubbles ... [in] the third type of display ..., triggered by some form of disturbance, generally directed from the front of the moth. ... The moth walks forward, flapping its wings[,] ... the wing movements being accompanied by a distinctly audible rattle[,] ... and the defensive fluid [is] secreted.

Most students of the subject have assumed that the multiple warning signals of an individual are all aimed at the same class of predators, such as day-flying birds that forage for insects on plants, and that these signals act synergistically, reinforcing one another. But John Ratcliffe and Marie Nydam argued—in an article aptly titled "Multimodal

Warning Signals for a Multiple Predator World"—that the separate signals may be aimed at different species of predators or at the same predators in different contexts. For example, at night, when visual signals are all but useless, a flying tiger moth informs bats of its chemical defense with sound, a string of largely ultrasonic clicks only partially audible to humans. But in daylight the very same moth rests quiet and motionless, often on the upper side of a leaf, its conspicuous warning colors informing diurnal birds of its unpalatability.

Warning signals benefit a well-defended animal in two ways. First, they may deter a predator from attacking, thereby sparing the insect (or other intended victim) the energy and time-consuming hassle of defending itself. Second, physical self-defense can be risky: the defender may be injured by the attacker. And if the insect's defense is toxic, it will deplete its store of "chemical warfare" agents, reducing its ability to defend itself—at least temporarily—against subsequent attackers. Furthermore, the energy and other resources required to produce these agents—which may be substantial—could be put to better use, to produce offspring that will pass the defender's genes on to future generations.

Now we come to the seeming paradox of how an insect can benefit from being eaten by a predator. Remember that the antipredator weapons of insects, including toxins that have an effect only if they are eaten, are generally not deadly. The intention is almost always not to kill predators but to educate them. Consequently, most if not all predators in an area soon learn not to eat insects of the kind—usually conspicuously colored—that made them sick, thereby making the area safer for relatives of the victims that were eaten. Even a dead victim can benefit, as William Hamilton's brilliant concept of kin selection explains. One way to measure an organism's evolutionary fitness is to count how many offspring, the bearers of its genes, it leaves behind. Hamilton proposed that an individual eaten by a predator has "sacrificed itself" for the good of its relatives, which contain many of the victim's genes. He argued that the ultimate measure of an individual's evolutionary success is its *inclusive* fitness, the survival of its genes in the bodies of its own offspring or those of its relatives. Each parent gives its offspring half of its genes: thus sister and brother share half of their genes, first cousins share one-eighth of their genes, and one-quarter of a grandparent's genes are passed on to each grandchild.

Insects and other arthropods are not the only aposematic animals. For example, the four species of North American skunks are among the relatively small number of aposematic mammals. Almost all North American small mammals—raccoons, rabbits, mice, squirrels—are an inconspicuous gray or brown, but, as we see in figure 8, the skunks are flagrantly conspicuous in shining black and white, advertising their potent chemical defense. The striped skunk, common throughout southern Canada and the United States, warns unwelcome intruders, as Donald Hoffmeister and Carl Mohr described, "by elevating its tail straight as an exclamation mark, pluming out its tail hair, and stamping its front feet." If the intruder ignores this warning and comes too close, the skunk, its back turned to its target, takes aim and sprays the unwelcome visitor with the malodorous and irritating contents of its scent glands.

Of the many insects that feed on milkweeds, almost all—among them adult monarchs and several species of true bugs, aphids, and beetles—are variously patterned with red and black. Nearly 90 percent of the stinging wasps of the family Vespidae—the potter wasps, paper wasps, yellow jackets, and hornets—are dark in color, usually black, and boldly marked with bands of bright yellow or occasionally white. Many of the insectivorous ladybird beetles are red with black spots or black with red spots, a warning of the noxious alkaloids they make and store in their bodies. A German zoologist, Fritz Müller, noticed that, similarly, various species of South American butterflies that all feed on toxic plants are so alike in appearance that they can barely be told apart. In 1879 he published an explanation of why well-defended insects of different species and even different families or orders may so closely resemble one another, a phenomenon now referred to as Müllerian mimicry and known to be widespread and fairly common.

Müller postulated that noxious and warningly colored species converge on a common signal, through natural selection, because they benefit from adopting similar "advertising logos." Because some individuals of even the most toxic species will be killed in the process of educating and reeducating predators, he reasoned that if two or more species have the same or similar warning signals, the members of all of these species will benefit. There is only one signal for predators to learn, and because they can't or find it difficult to tell these species apart, the inevitable mortality will be shared by a larger number of individuals.

Numerous experiments and observations—most done in the laboratory and a few in the field—have shown that birds and other predators are deterred by chemical defenses and can learn to pass up conspicuous, unpalatable prey. Tim Guilford lists several publications that offer evidence indicating that one or more species of toads, lizards, birds, and fish can learn this lesson. As we will see below, even a "lowly" insect, a praying mantis, is capable of learning not even to try to eat a noxious insect.

Thomas Boyden did an exceptionally interesting and convincing field experiment with lizards (genus *Ameiva*) in Panama. By casting with a fishing rod, he presented wild, free-ranging *Ameivas* with palatable, living nonaposematic butterflies (*Anartia fatima*) and blatantly aposematic ones, the red postman (*Heliconius erato*). Although tethered, the butterflies were able "to fly freely in a normal fashion." The *Heliconius* were known to repulse birds because they sequester cyanide compounds from their food plants, passionflowers. Of nineteen palatable *Anartias* cast to them, the lizards attacked 95 percent, killed 89 percent, and ate 84 percent. On the other hand, they were obviously repulsed by the noxious *Heliconius*, attacking only 47 percent, killing only 22 percent, and eating none. Another of Boyden's experiments leaves no doubt that it was *Heliconius*'s conspicuousness that warned away the lizards. If the butterflies' warning colors were covered with black paint, the lizards attacked 100 percent of them—but presumably because they sensed the cyanide compounds, the lizards killed only 33 percent, and ate none.

In 1969, Lincoln Brower, whom I think of as Dr. Mimicry, did an exceptionally clever and convincing laboratory experiment with blue jays and monarch butterflies. He caged wild-caught blue jays until they had forgotten any previous encounters they had had with monarchs in the wild. Then he offered the "brainwashed" birds monarchs that were palatable because they had been raised in the laboratory on one of the species of milkweed that contain no cardiac glycosides. The jays liked the monarchs, which they continued to eat until after Brower fed them unpalatable monarchs that had been raised on a species of milkweed loaded with cardiac glycosides. Jays that ate these noxious monarchs quickly showed signs of distress, raising their crests and fluffing out their feathers. They then vomited, as often as nine times in half an hour. This was one-shot learning. Thereafter, the jays would not even touch a monarch, either palatable or noxious, and some even retched just at the sight of one.

Praying mantises will eat almost anything that moves—usually insects, including even other mantises—and there have been a few reports of one snatching a hummingbird hovering at a sugar-water feeder. But until May Berenbaum and Eugene Miliczky did some experiments with the introduced Chinese mantis (*Tenodera sinensis*), little was known about how mantises are affected by toxic insects and whether or not they could learn not to eat them. Mantises, Berenbaum and Miliczky observed, grasped milkweed bugs with their forelegs without hesitation and began to devour them but soon tossed them away and shook their forelegs vigorously. Then many vomited large quantities of an orange liquid. As we have seen, milkweed bugs seques-ter cardenolides from milkweed seeds and advertise their resulting unpalatability with their red and black warning colors. The memory of this sickening experience remained with the mantises for quite some time. Two rejected milkweed bugs three weeks after they had last seen one. They also rejected perfectly edible beetles that had been painted red and black. Some years later, Todd Bowdish and Thomas Bultman showed that mantises promptly captured and ate milkweed bugs that were palatable because they had been raised on sunflower seeds, which do not contain cardenolides. Bowdish and Bultman also found that the previously "educated" mantises were less likely to strike at milkweed bugs whose natural color pattern had not been altered or those that had been painted with alternating stripes of orange and black than at others that had been painted all orange or all black.

There have been relatively few studies of the responses of predaceous insects to noxious and warningly colored prey, although Bowdish and Bultman pointed out that "there is accumulating evidence for the ability of insects and other invertebrates to learn" from experience in other contexts. The most widely known example is worker honeybees' learning the route and distance to a patch of nectar flowers by dancing in the hive alongside scout bees doing the waggle dance, which indicates the direction and distance to the flowers.

# NINE

# The Predators' Countermeasures

As we saw in chapter 2, insectivores have evolved over the eons and are still evolving new and sometimes bizarre adaptations and tactics to not only counter the evolving strategies of defensive insects but also minimize competition from rival insectivores. A few birds, as we will see below, feed on toxic monarch butterflies by ingesting only the least toxic parts of their bodies and discarding the rest, while other birds disarm stinging wasps and bees by beating them against a perch until the abdomen, which bears the sting, falls off. Soaring swallows and chimney swifts both pursue flying insects but improve their hunting success by dividing the airspace to minimize competition. Swallows stay close to the ground, and the swifts fly higher, above the roofs of buildings in cities, where they earned their name by gluing their twig nests to the inside walls of chimneys. Foraging brown creepers and nuthatches, at least to some extent, share the insects, insect eggs, and other tiny creatures tucked away in bark crevices. Nuthatches, which often creep headfirst *down* a tree trunk, are likely to spot insects missed by brown creepers, which always climb *up* the trunk. Insectivorous birds of many other species have evolved tactics for improving their odds as they search for hidden or camouflaged prey. The black-capped chickadee, as we will see, uses a particularly amazing tactic.

Figure 9. This eastern kingbird, a New World flycatcher, has
snatched a honeybee from the air. Like many other flycatchers,
it can recognize stingless drones and pass up stinging workers.

Monarch butterflies that breed east of the Rocky Mountains in Canada
and the United States make a long migration to spend the winter in
oyamel fir forests high in the mountains of Mexico. Millions of them
crowd into thirteen known sites of only a few acres each, so densely
crammed together that they obscure the foliage of the trees on which
they cluster.

Only one kind of mouse and two of several dozen insectivorous bird species in the vicinity of the overwintering sites, according to William Calvert and his coauthors, manage to make significant use of this immense bonanza of food without being sickened by the monarchs' cardenolides. Nevertheless, a few other mice and birds, probably exceptionally daring or hungry individuals, occasionally eat or try to eat a monarch.

The two birds that regularly eat monarchs, black-headed grosbeaks (*Pheuticus melanocephalus*) and black-backed orioles (*Icterus galbula*), have evolved two different safe ways of ingesting these often toxic butterflies (some are not toxic because they contain no cardenolides). The grosbeaks can eat monarchs because they are physiologically tolerant of cardenolides. After discarding the nutrient-deficient wings, as do most if not all birds that eat moths and butterflies, the grosbeak snaps off the monarch's abdomen with its big, powerful beak, swallows it whole, and then eats some of the muscles and other tissues in the thorax. But the orioles are not immune to or tolerant of cardenolides, and their way of feeding on monarchs is consequently selective. "Orioles kill monarchs randomly with respect to cardenolide content," Linda Fink and Lincoln Brower explained, "but then proceed to eat them in ways that reduce the amount of cardenolide that is ingested." With their long, thin beaks, they slit and pry open the abdomen and the thorax and withdraw and eat only the internal muscles and other organs, which have the lowest concentration of cardenolides. The exoskeleton, the body wall, which has a high content of cardenolides, is left intact and discarded, just as we discard a lobster's hard shell.

Black-eared mice (*Peromyscus melanotis*) eat both dead and living monarchs that have fallen to the ground and living ones perched on low vegetation, John Glendinning, Alfonso Mejia, and Lincoln Brower reported. These mice often eat the entire abdomen of dead, dried-out monarchs but, like the orioles, feed selectively on the abdomens of living and still fresh and moist dead ones "by discarding the bitter, cardenolide-laden [exoskeleton] and eating the internal tissues."

During the 135-day overwintering season, the mice kill as many as 570,000 monarchs per hectare (2.5 acres); that's a lot of butterflies but only about 5 percent of the estimated ten million per hectare.

The grosbeaks and the orioles eat a great many more monarchs: more than two million during an overwintering season in just one of the five Mexican sites investigated by Calvert and his colleagues. This is, however, only about 9 percent of the more than twenty-two million monarchs that occupied this site. Assuming that the mice also ate 5 percent, as Glendinning and his coauthors showed they did in a later year, the total loss of monarchs to predators in the earlier winter season was about 14 percent, about three million, of those present. This left well over nineteen million to move northward and on the way produce progeny that return to the United States and Canada to breed a new generation or generations of monarchs, the last one of which will make the long southward migration and spend the next winter in Mexico.

"Know your enemy" is a fundamental military axiom. Reworded as "Know your prey," it is good advice for lizards, birds, mammals, and other insect eaters. The observations of the beekeeper that I quoted in chapter 2 aptly illustrate the value of this advice. He discovered that the Arizona crested flycatcher eats stingless drones to avoid being stung by female honeybees. I have seen no confirming evidence in the literature but do not doubt that his astute observations are accurate. They raise two interesting questions: how do the flycatchers know when drones are numerous, and how do they tell a drone from a similar worker bee? The answer to the first question is most likely that these desert-dwelling birds, which have to explore far and wide for food, just happen to come upon an apiary at the right time, when drones are abundant. Second, they almost certainly recognize drones by their larger size and somewhat different shape, differences that are probably obvious to the birds—but not to most people—when they see drones flying amid a crowd of smaller workers. According to Arthur Cleveland Bent, other flycatchers also attack drones and pass up the fiercely stinging workers.

Of the fifty honeybees in the stomachs of several eastern kingbirds, he reported, forty were drones, six were too damaged to determine their sex, and only four were workers. The stomachs of five western king-birds, a related species, contained twenty-nine dead drones and only two workers.

The perceptive Arizona beekeeper noticed that summer tanagers also feast on honeybees. They perch in trees that shade the hive, and pounce on returning worker bees that slow down to land at the entrance to the hive. He described to Herbert Brandt how this beautiful bird—although females have subdued colors, the magnificent males are virtu-ally all bright red—captures its prey: "Its food in the areas about my apiaries consisted almost entirely of bees, and worker bees, at that. Or I would better say, parts of bees, for the bird skillfully avoided contact with the stinger end of its victim by breaking off that end. This was accomplished by catching the bee across the middle of the body and, on alighting on a branch or other perch, break[ing] off the protruding end of the abdomen by giving it a sharp sweep across the perch." He went on to say that "evidently the several pairs of . . . tanagers that reared their young in the vicinity of my apiaries, fed mostly on bees. . . . All summer long the top of nearly every hive was sprinkled with the abdo-mens of bees, and just as many probably fell on the . . . ground through-out the apiary."

Other birds also disarm stinging insects in one way or another, among them western tanagers, certain flycatchers, shrikes, and bee eaters. The bee eaters, a small family (Meropidae) of fewer than thirty species distantly related to the kingfishers, occur only in the Old World. My daughter Susan and I watched a blue-tailed bee eater (*Merops philippinus*) perched on a telephone wire on the island of Luzon in the Philippines. It was large and colorful and had a long forked tail and a long thin beak. We were struck by its behavior. Like one of our North American flycatchers, it sallied out to snatch insects from the air. We didn't get a good look at the insects, but they were most likely bees or wasps, judging by Malcolm Edmunds's report that the prey of eleven other bee eaters includes from more than 60 to more than 90 percent stinging insects. The bee eater immediately returned to its perch and painstakingly manipulated its catch before swallowing it. We could not see exactly what it did, but Edmunds noted that these birds "bang the head of the [stinging] insect against the perch, and bang and rub the abdomen several times to squeeze the venom out of the sting gland. They give a final one or two head blows, and then swallow the insect head first.... Non-venomous insects are simply beaten on the head before swallowing—the abdomen is not rubbed."

Many insects hide from their enemies, but predators, notably birds, have many different ways of finding them. Some insects, such as wood borers, will be out of sight, at least in the larval stage, because of their lifestyle. Other insects are camouflaged and hide in plain sight; yet others conceal themselves, for example, in the foliage of trees or in fields of grass and other low vegetation.

Some birds take advantage of inadvertent beaters to flush well-concealed insects. "In tropical America," Roger Tory Peterson noted, "a number of soberly colored ant birds specialize in following the large swarms of army ants, feeding on the many insects that are flushed up as the army advances over the jungle floor." In eastern Africa, cattle egrets, bee eaters, and other birds follow antelopes, elephants, zebras, rhinoceroses, and other grazers to catch the moths, grasshoppers, and other insects that they flush as they move about cropping grass. Similarly, in North America the aptly named brown-headed cowbirds follow cattle, just as they once followed the immense herds of bison that wandered the plains, feeding on the insects they flushed. (Cowbirds still lay their eggs in the nests of other birds, which, on the plains, would have stayed behind when the bison moved on too soon for the cowbirds to complete raising a brood of young.)

More than one hundred years ago, cattle egrets crossed the Atlantic from Africa to the New World, apparently without human assistance. Away from the many grazers that inadvertently served them in Africa,

North American egrets associate with other flushers, mainly cows and farm machinery. Harold Heatwole found that this greatly benefits these birds. Watching two cattle egrets in the same pasture, he and an assistant found that, while taking only two-thirds as many steps, the one following a cow captured 25 to 50 percent more prey than the one foraging alone.

A cocoon protects many insects, most famously moth caterpillars, during the pupal stage, when they are especially vulnerable to predators because being virtually immobile, they cannot run away or defend themselves. As John Henry Comstock put it, before molting to that stage, caterpillars "make provision for this helpless period by spinning a silken armor about their bodies." Among the moths, several of the giant silkworms (family Saturniidae) of North America spin large, tough-walled cocoons in which they spend the winter in the pupal stage. The huge cecropia caterpillar, for instance, constructs a double-walled cocoon 3 or more inches long and immovably attached along its length to a sturdy twig. Most cecropia caterpillars descend from the trees they have fed on to attach their cocoons to the stems of shrubs near ground level, where they are less likely to be found by predators.

During several winters, James Sternburg and I collected about three thousand cecropia cocoons from trees and shrubs in the twin cities of Champaign and Urbana, Illinois. Predators had killed many of the pupae by removing their soft tissues through a tiny hole punched through the cocoon wall. More than 86 percent of cocoons in trees had been attacked versus fewer than 19 percent of those spun by caterpillars in shrubs near ground level.

We thought that the predator had to be a woodpecker, but science demands more than an educated guess. After many months we finally caught a wild downy woodpecker in the act. Before then, William George, an ornithologist at Eastern Illinois University, had obliged us by putting a cocoon in a cage with a hand-reared hairy woodpecker that had never before seen a cocoon of any kind. The bird soon pierced the cocoon with its chisel-like bill and lapped up the pupa's soft tissues with its long, extensible barbed tongue. George, wondering why the woodpecker had thought there was something to eat in the cocoon, reasoned that these birds must be opportunists that check out unfamiliar objects. He was right. He found that in a woodland, unusual objects, such as healed wounds on branches or burls on tree trunks, were always marked by the pecking of woodpeckers.

Although cecropia moths were then abundant in the Twin Cities, they were decidedly uncommon in rural areas. Aubrey Scarbrough, Sternburg, and I found out why. In the country, cocoons well above ground level were attacked by woodpeckers, just as in the cities, but those spun near ground level, although seldom found by woodpeckers, were not safe from other predators. Mice had torn large holes in and removed the pupa from more than 57 percent of them. Another student in the University of Illinois entomology department found three intact cocoons cached next to a mouse's nest under a log. In the cities, on the other hand, only three of three thousand cocoons had been attacked by mice. Why this big difference?

We set live and snap traps outdoors and caught ninety-one mice in rural areas, eighty-four of them native white-footed or deer mice (*Peromyscus*) and only seven nonnative house mice *(Mus)*. Traps placed under shrubs in Champaign-Urbana caught ninety-five house mice but only seven native mice. We acclimated both native mice and house mice caught in live traps to cages, then presented them with cecropia cocoons, toward which they behaved very differently. The native mice readily opened the cocoons and ate the pupae. All of the house mice ate pupae we had removed from cocoons, but not one ever tore open a cocoon to get a pupa, although all ripped off large amounts of silk, which they incorporated in their nests.

Architecturally, the cocoon of another giant American silkworm, promethea, is radically different from that of the cecropia. Instead of being immovably attached to a twig, it dangles by a thin, flexible silk strap about an inch long from a slender twig, usually at the end of a branch, high in a wild cherry or sassafras tree. Woodpeckers seldom penetrate these cocoons, apparently because they swing away when the birds peck at them. Of 412 promethea cocoons Sternburg and I collected in nature, only 13 had been penetrated by a woodpecker. (We once found a few promethea cocoons partly exposed but still firmly embedded in a melting snowbank. The ever-opportunistic woodpeckers had found and penetrated all of them.) Only 2 of the 412 cocoons had been attacked by mice, presumably because, fearing predators, they are loath to climb high in a tree, although they will climb onto its lower branches.

John Alcock noted that the Dutch animal behaviorist Luuk Tinbergen, the brother of Niko Tinbergen, was the first to suggest that "insect-eating birds might learn from experience to search specifically for the subtle visual characters that give away the presence of cryptic prey. This, the search image hypothesis, proposes that predators can better locate well-concealed but profitable prey by scanning for the particular cues associated with a particular species." Since Tinbergen advanced this hypothesis in 1960, many experiments, several of which Alcock summarized, have supported it.

The formation of a search image by birds, Malcolm Edmunds argued, is the principal threat to a camouflaged insect. This danger is lessened "if the prey species is rarely encountered, as the predator will soon forget the searching image. So cryptic species must be spaced widely if they are to gain full protection from their primary defence." Among the examples he gives of insects that practice such spacing is the mourning cloak butterfly. The warningly colored caterpillars are gregarious, "but the pupae are cryptic and just before pupation [the caterpillars] crawl off in different directions to pupate in isolation." A revealing example cited by Edmunds is a fish that "deposits spawn in compact masses if there are no predators around, but scatters it when predators are present." As discussed in chapter 7, the article "An Experiment on Spacing-Out as a Defence against Predation" by Niko Tinbergen and his students described the behavior of wild carrion crows searching for hens' eggs painted with a camouflaging pattern. Tinbergen and his students spaced the eggs closely or widely and found that wide spacing really is advantageous for camouflaged prey animals.

What we have seen so far leaves no doubt that insects of a great many species are not easy to find, because they are hidden or camouflaged and, in both cases, usually scattered and widely separated from one another. Consequently, most insectivores spend a lot of time—often all or much of the day—searching for insects. To increase their efficiency—how often they catch an insect—many have, mainly via natural selection, refined their hunting techniques, and a few even hunt together with insectivores of other species for reasons that I hope to make clear next.

Bird watchers and other naturalists sometimes see, especially in winter when the trees are bare, mixed feeding flocks consisting of several species of birds moving through the woods as they flit through trees and shrubs hunting for insects. How and if birds benefit by joining a mixed flock was a subject for speculation until Kimberly Sullivan studied the foraging flocks of chickadees, tufted titmice, downy woodpeckers, red-breasted and white-breasted nuthatches, brown creepers, and occasionally other insect-eating birds that commonly occur in southern Canada and the eastern United States. She demonstrated that downy woodpeckers—and presumably the other birds—in mixed flocks benefit because they can devote less time to watching for predators and more time to searching for food and eating than when they are alone.

The birds in these flocks, and probably almost all other birds, vocalize, in addition to their songs, at least two kinds of calls: frequent social calls characteristic of their species, which inform one another that they are present and vigilant, and loud alarm calls understood by most if not all other kinds of birds and even mammals, which warn of the approach of a sharp-shinned hawk or other bird-eating predator.

Using this knowledge and her observation that foraging downy woodpeckers frequently cock their heads as if to watch for predators,

Sullivan designed a field experiment that showed that downies cock their heads more often when foraging alone than when they hear the constant social calls of their flockmates. She played the recorded social calls of black-capped chickadees and tufted titmice to lone downies foraging in the wild. Presumably thinking that flockmates were with them and watching for predators, these lone woodpeckers cocked their heads less often and spent more time feeding than when they, although other birds were not actually nearby, were serenaded by Sullivan's recording. In control experiments, lone foraging downies did not cock their heads less often in response to recorded social calls of seed-eating birds— American tree sparrows, goldfinches, and juncos—none of which ever flock with them.

Black-capped chickadees eat a wide variety of both animal and vegetable matter and are always on the lookout for anything edible. "Chickadees," Susan Smith explained in her book on these birds, "are strongly inquisitive, tending to explore any new material they encounter." In a winter woodland in Connecticut, I watched a black-capped chickadee hanging upside-down from a cecropia cocoon, scraping at the silk with its beak until it had cut a long slit and then plunging its beak into the soft tissues of the pupa. Smith described other chickadee foraging behaviors, including how a group pecked through the skin of a dead deer to get at the subcutaneous fat and how one bird satisfied its sweet tooth by knocking a small icicle from a sugar maple tree, catching it in midair, perching with it on a branch, and then eating all of it. However, I think that the most remarkable of their tactics is the one for finding caterpillars that Bernd Heinrich discovered and described.

These spry little birds are agile acrobats that nimbly hop from twig to twig and often hang upside-down as they inspect a leaf for insects. But their skills are not only athletic. They display what—at least in my view—can only be called intelligence as they search for their prey. Heinrich explained that the hunting behavior of these charming little birds is amazingly clever and sophisticated. Chickadees keep an eye out for leaves—those that are tattered or holey—that have been partially eaten, recognizing that these damaged leaves indicate that caterpillars are probably on them or on nearby leaves. Heinrich and Scott Collins discovered this unexpected behavior by observing captive chickadees in large screened outdoor cages that enclosed large vertical branches serving as stand-ins for trees. The researchers artificially damaged some leaves and placed pieces of succulent beetle grubs nearby as stand-ins for caterpillars. The birds soon learned that a damaged leaf indicated the presence of a meal. In another experiment, chickadees were given a number of opportunities (foraging trials) to search for food in an aviary containing ten small trees, only two of which had damaged leaves with food nearby. "After about ten foraging trials the chickadees no longer searched ... on the trees with undamaged leaves. Instead they flew straight to one of the two trees with damaged leaves." Heinrich and Collins confirmed their results by observing wild chickadees in the field.

They also discovered that camouflaged caterpillars trim leaves to minimize the conspicuous damage from their feeding and even completely remove from the tree leaves upon which they had fed, clues that could alert birds to their presence. The twenty-six species of caterpillars Heinrich and Collins observed sometimes trimmed tattered or otherwise damaged areas from a leaf before moving on to another. All also had an alternative tactic. Six species always ate the whole leaf; sixteen only sometimes ate the whole leaf but always otherwise clipped off the uneaten remnant, which fell to the ground after they severed the stem; four species always ate only part of the leaf and always clipped off the remnant. White underwing caterpillars sometimes trimmed partially eaten leaves, sometimes ate all of a leaf, and sometimes clipped off partially eaten leaves.

In an earlier article, Heinrich described how certain sphinx caterpillars, tobacco hornworms (*Manduca sexta*), manage to eat all of a large, broad tobacco leaf that is much bigger than they are. They firmly grasp the stem of the plant near the base of the leaf with the fleshy, stubby "prolegs" at the hind end of the long body and, as they consume the leaf at its edge, pull the leaf toward themselves, causing it to bend, by "walking" forward on the leaf using the legs on the thorax at the front end of the body. As they near its tip, they continue to consume the bowed leaf by chewing back toward its base. They repeat this maneuver until they have consumed the entire leaf.

The caterpillars whose feeding behavior I have been discussing are all palatable, luscious morsels for a bird. As is to be expected, all are camouflaged, hiding by blending in with the background. As we have seen, unpalatable caterpillars are warningly colored. Birds learn to recognize these warnings and do not eat these insects. Conspicuously warningly colored caterpillars, Heinrich and Collins reasoned, should have no need to eliminate the signs of their feeding. And indeed, not one of the thirteen species of warningly colored caterpillars they observed made any attempt to disguise or eliminate damaged leaves. It is as if they were saying to themselves, "Why should I bother? My bright colors will scare off predators anyway."

Most moths and many other adult insects lie low during the day and, in order to minimize their exposure to diurnal predators, fly only at night. On summer nights the air is often crowded with flying insects, as demonstrated by the swarms of moths, beetles, and other insects attracted to lighted windows. There are sometimes vast clouds of night-flying insects that we don't see, but there is no doubt that they exist, as witnessed by the fact that vast hordes of bats, like those at Carlsbad Caverns National Park in New Mexico, obviously catch enough insects every night, or at least on most nights, to sustain themselves and their young. David Roemer, a biologist at the park, told me that almost nine million Mexican free-tailed bats roosted in the caverns in 1936, before DDT poisoning exterminated most of them. These bats, mostly females raising a rapidly growing baby, in one night ate about 1.16 million pounds, 58 tons, of flying insects, mostly small moths and beetles. That is a lot of insects. Since these moths and beetles weigh only a tiny fraction of a pound each, in that one night the bats, by my rough estimate, had to catch and eat somewhere between half a billion and one billion insects to satisfy their appetites. Nevertheless, millions of moths, many of them crop pests, survived, enough to continue their northward migration and repopulate the Midwestern Corn Belt.

This bonanza of flying insects is exploited by nightjars such as whip-poor-wills and nighthawks and owls such as the eastern, western, and whiskered screech owls, but mainly by various species of bats, the undisputed masters of the night skies. They are the only animals other than insects and birds that can fly, not just glide like flying fish and flying squirrels. All bats are nocturnal, and most feed only or mainly on insects, as do all forty-five North American species.

How does a flying bat find a flying insect at night? Bats are not blind, but vision is not the only way—and perhaps no way at all—to accurately locate a flying insect in the dark and plot a course to intercept it. Bats detect and track their prey by means of sonar, or echolocation, familiar from old movies about submarines. In a tense scene, sailors on the submerged vessel hear a slow *ping, ping, ping,* the sound waves of a destroyer's sonar probing for an enemy submarine. Returning echoes that bounce off the sub's hull reveal its depth and location.

According to Wilfried Shober, bats "invented" sonar long before we did, about fifty million years ago. Before the eighteenth century, bats' ability to navigate and catch insects in the dark was attributed to magical powers. At the end of the eighteenth century Lazzaro Spallanzani found that covering or even removing a bat's eyes did not affect its ability to navigate in the dark, and the Swiss zoologist Charles Jurine demonstrated that bats could not navigate if their ears were plugged. There was then no doubt that bats need their sense of hearing to navigate. But it was not until the 1950s that Donald Griffin firmly established that bats both navigate and locate insects by echolocation. In his delightful and readable *Listening in the Dark,* Griffin discussed bats and the fruit-eating South American oil birds, which spend the day in caves and navigate by sonar at night; and the East Indian swiftlets, some of bird's-nest soup fame, which chase flying insects by day, like the related North American chimney swift, and use sonar only to navigate in the dark caves in which they nest and rest at night.

Shober gave a simple and concise explanation of how bats use echolocation:

> When bats fly at night, they emit short, regular sound pulses. The echoes that are bounced back provide the bats with information about obstacles in their flight path and about insects that are flying nearby. As soon as an insect is heard or enters the sound-sensing beam of a bat, the frequency of impulses is increased instantly to allow its exact location and pursuit. The noise produced, although inaudible to the human ear [because it is too high pitched] registers an intensity greater than that of a pneumatic hammer.

It has been known for a long time that some moths have ears. Until about fifty years ago this was quite a puzzle for entomologists because they assumed that moths do not themselves make sounds. (We now know that a few do.) What, scientists wondered, do moths listen to if not one another? These night-flying insects, as you've probably figured out, are on the alert for the sonar cries of bats, which warn them that a predator is close by and will catch and eat them if they do not take evasive action.

Kenneth Roeder, an insect physiologist at Tufts University, happened to notice that some of the moths circling a streetlight made apparently evasive maneuvers when bats approached. Others had made similar observations, as summarized by Asher Treat, and as early as 1925, German investigators had shown that some moths "are sensitive to airborne vibrations of high or ultrasonic frequency." But the connection between moth ears and bats was not made until the 1940s and 1950s, after Griffin discovered that bats avoid obstacles and locate insects by means of ultrasonic echolocation.

"Moths," Roeder wrote in *Nerve Cells and Insect Behavior,* "are one of the main food sources of certain families of bats. They are attacked on the wing and in darkness in a contest in which speed and maneuverability are the premium qualities." Moths can detect a bat cry at a distance of 100 feet and perhaps more. A distant cry is of low intensity, and the moths will move unhurriedly away from its point of origin, out of the area in which the bat is hunting. If the bat is close, the cry is of high intensity, and a panicked moth will make sudden evasive maneuvers: "dives, rolls, repeated tight turns, or rapid flight above the ground." These maneuvers often save the moth—"for every 100 reacting moths that survived attack only 60 nonreacting moths survived."

Roeder studied owlet moths (family Noctuidae), but other insects also have ears that respond to ultrasound. Some have been shown to listen for bats, and the others presumably do the same. Like the owlet moths, measuring worm moths (family Geometridae) and tiger moths (family Erebidae) have ears, tympanic organs, one on either side of the base of the thorax. On a warm Texas night, a guest at a cocktail party asked Roeder if the sphinx moths (family Sphingidae) hovering at the flowers near a patio have ears. He said he didn't know and jingled his keys, which produced some ultrasound as well as audible noise. Instantly, to his surprise, the moths took evasive action. Back at his laboratory at Tufts, he spent many hours looking for the sphinx moth's ears. He eventually found them in an unexpected location, one on each of two appendages near the mouth.

Nocturnal flying insects other than moths, Mike May reported, have ears and respond to artificial ultrasonic pulses that mimic the cries of bats: green lacewings (family Chrysopidae) have an ear on each forewing; some praying mantises have a single ear between their hind legs; locusts, like other grasshoppers, bear an ear on either side of the first abdominal segment; katydids and crickets have an ear on each front leg; and tiger beetles have a pair on the upper side of the first abdominal segment. The location of the ears of these insects on seven different parts of the body leaves no doubt that evolution, via natural selection, "invented" the same strategy for escaping from bats at least seven times.

When alerted by a bat's cry, owlet moths, lacewings, and other flying insects try to escape by making evasive maneuvers, or, like cockroaches, houseflies, and most other insects, they flee. But tiger moths do not make evasive maneuvers or flee when threatened while flying at night. They—it can be argued—defend themselves against bats the way monarchs and other unpalatable, aposematic insects defend themselves against birds. Tiger moths are definitely unpalatable, probably to most insectivores. (When my friend and colleague Bill Downes offered tiger moths to his pet bats as I watched, they always rejected these insects with obvious disgust.) Relying on their warning coloration to keep them safe, many tiger moths that are inactive during the day are conspicuous as they rest motionless in plain sight, often on the upper side of a leaf. This warning deters birds, but when these moths fly at night it is invisible to bats. Only a vocal signal could warn bats of their noxiousness from a safe—for the moths—distance. Some tiger moths can, in fact, make ultrasonic clicking sounds.

Dorothy Dunning and Roeder showed that bats intercepted and ate beetle larvae tossed into the air but turned away from these delicious tidbits when the recorded ultrasonic sounds made by an arctiid moth were played. They suggested that the moths' sounds may protect them from bats but took no stand on how the moths' sounds might affect the bats. A few years later, Dunning published evidence that "these signals can operate as 'aposematic sounds,' comparable to warning coloration."

Another possibility is that the moths' sounds jam the bats' sonar. "To distract an approaching bat that is about to try to catch it," James Fullard and his coauthors argued, "the moth produces clicks with acoustical characteristics closely resembling the returning echoes of the calls produced by many bats in the terminal stages of their approach to a target. The moth sounds [are] ... probably ... similar enough to real echoes to disrupt information processing by the bat."

I think that the moths' sounds are almost surely aposematic warnings. In Dunning and Roeder's experiments, bats serenaded by tiger moth sounds dodged away from tossed beetle larvae an average of about 85 percent of the time. When serenaded with the recorded cries of their own species, however, the bats dodged away only about 14 percent of the time, seemingly contradicting the argument that the bats' sonar is jammed by the moths' sounds.

Not until recently was the sonar jamming–aposematic warning controversy laid to rest. In 2005, Nikolay Hristov and William Conner reported the results of experiments comparing the response of naive big brown bats (*Eptesicus fuscus*) to three groups of the same species of milkweed-eating moths (*Euchaeteas egle*) with different combinations of palatability and muteness. In nature this moth responds to bat cries and is palatable because it feeds on a milkweed with a meager concentration of cardenolides. Hristov and Conner made some moths unpalatable by raising them on a cardenolide-rich milkweed and silenced some by impairing their sound-producing organs. Bats kept in a large enclosure greedily ate moths that were palatable but still capable of producing ultrasonic sound. However, they quickly learned to avoid—some after only one trial—moths that were *both* unpalatable and able to make sound. "The results of these experiments," Hristov and Conner concluded, "consistently point to acoustic aposematism as the raison d'être of sound production in species of tiger moths that we studied." Aaron Corcoran and his coauthors (including Conner) later published convincing evidence that clicking sounds made by moths can jam a bat's sonar. As they pointed out, however, jamming sonar and warning of unpalatability are not mutually exclusive.

As we have seen, predators have evolved many ways of getting around insects' defenses. But the escalating arms race between predators and prey continues and will go on as long as there is life on earth. Some insects and other animals, including even a few snakes, have adopted an amazing defensive tactic. As we will see in the next chapter, predators pass up many perfectly palatable and harmless insects because they look like insects that are unpalatable or armed with painful stings.

# TEN

# Protection by Deception

One of the best ways not to be eaten is to resemble something that isn't good to eat. In South Africa there are plants that escape the notice of grazers because they look like rocks lying on the ground. Many animals, most of them insects, mimic other animals that bite, sting, or are unpalatable because they contain a sickening chemical substance. Nonvenomous snakes mimic the deadly coral snakes; many harmless insects—flies, moths, and even beetles—mimic stinging bees and wasps; the viceroy butterfly mimics the toxic monarch butterfly. There are even, improbable as it seems, caterpillars that mimic venomous snakes.

Most of the more than four thousand known species of cockroaches—only a few dozen are household pests—are camouflaged, secretive, and nocturnal. But improbable as it seems, in the tropical Philippine Islands there are some atypical cockroaches (genus *Prosoplecta*) that mimic unpalatable ladybird or leaf beetles. They are active in daylight, not secretive, and as conspicuously colored as their models. When I first saw the beautiful pictures of these cockroaches in Wolfgang Wickler's book on mimicry, I was flabbergasted. I had probably assumed, perhaps subconsciously, that all cockroaches are mostly dull brown and tan, that colorful mimetic cockroaches do not exist. (I had known that some tropical cockroaches are green, blending in with

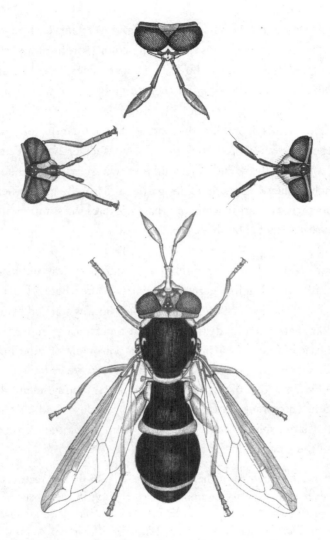

Figure 10. Hoverflies have short, three-segmented antennae, while wasps have long multisegmented antennae.
Three of the wasp-mimicking hoverflies shown here
(clockwise from the top: *Ceriana signifera, Sphecomyia vittata,
Tenthredomyia abbreviata*) have different ways of lengthening
their antennae without adding an additional segment, but
the one (*Spilomyia hamifera*) on the left waves its front legs to
simulate wasps' antennae.

the plants on which they live.) According to Wickler, each of several Philippine cockroach species mimics the color pattern of a particular species of red and black ladybird. The cockroaches, tasty morsels for lizards and birds, are protected by their resemblance to the unpalatable beetles. They also simulate the ladybirds' characteristic shape, short-bodied and hemispherical, unlike other predominantly long-winged and rather narrow-bodied cockroaches. As Wickler pointed out, the mimics are good flyers and have long hind wings, which, unlike other cockroaches, they compactly fold together and bundle under their colorful short wing covers in a manner that facilitates the simulation of the distinctive short, chubby shape of a ladybird.

The first report of a snake-mimicking caterpillar was published in 1862 by the great English naturalist Henry Walter Bates. During his extensive exploration of the Amazon Valley, he was startled by a large caterpillar that, when he disturbed it, did a persuasive impersonation of the head and neck of a small snake, "a poisonous or viperine spe-cies[,] . . . not an innocuous . . . snake." Almost sixty years later, the Reverend A. Miles Moss, the British chaplain of Pará (now Belem), Brazil, an amateur entomologist of professional caliber, described his encounter with a large sphinx moth caterpillar (family Sphingidae) near the mouth of the Amazon River:

> The larva is quite one of the most remarkable of living creatures that I have ever seen, a perfect Aaron's rod, combining in the most novel and striking way the principles of protective resemblance with an aggressive snake-mimicry. When at rest as an adult caterpillar, it hangs by two pairs of claspers in the vertical from the stem of its food-plant, and appears to be nothing but a broken branch covered with a dull, creamy white lichen. . . . The wonder, however, is if possible exceeded when, on being disturbed, this marvel of creative evolution endeavours once more to deceive by turn-ing into a snake. . . . The effect is produced by the creature turning itself over and exhibiting its ventral area, which is adorned by a broad band of dark olive-green with the three anterior sets of claspers completely with-drawn and scarcely visible. The thoracic segments, which are always

swollen, become puffed out laterally to an exaggerated extent; a pair of black eyes on segment 4, hitherto concealed and situated behind the now recumbent and wholly inconspicuous legs, open out: the cheeks appear to be adorned by yellow scales with black edges; and the fraudulent notion that one is beholding merely the head and neck of a formidable, if small, snake is carried to a nicety by the rigidity of the curve adopted. Then, as if to mesmerize, a swaying side-to-side motion is kept up for an appreciable number of seconds, before the creature, seeming to realize that an attack is not further contemplated, gradually closes its false eyes and relapses once more into diurnal slumbers.

Anecdotal evidence indicates that various insect-eating animals, including humans, are deterred by the snake-mimicking caterpillars of several sphinx moth species. The inhabitants of the village in which Bates lived were alarmed when he showed them his snake mimic. Hugh Cott cited observations by several naturalists: of sparrows, chaffinches, and chickens refusing to attack a displaying snakelike caterpillar; of lemurs, which are dedicated insectivores, terrified by one; of two captive baboons very much frightened by a snake mimic, the male screaming in "abject terror" when brought close to it.

Although extraordinary and uncommon, snake mimicry is only one of numerous examples—certainly of many thousands—of the mimicry of well-defended insects or other animals helping to protect innocuous edible animals from predators. In North America, according to the late John Bouseman and James Sternburg, the spice bush swallowtail (*Papilio troilus*, family Papilionidae) "resembles the head of the arboreal rough green snake." But I think it's a less convincing snake mimic than are the South American sphinx moth caterpillars.

A few animals besides insects—some snakes, birds, salamanders, mammals, and spiders—are also defensive mimics. This type of mimicry was first investigated and understood by Henry Walter Bates, who said, "When we see a species of Moth which frequents flowers in the daytime wearing the appearance of a wasp, we feel compelled to infer that the imitation is intended to protect the otherwise defenceless insect by deceiving insectivorous animals, which persecute the Moth, but avoid the Wasp." In honor of its discoverer this form of mimicry is now known as Batesian mimicry.

Batesian mimicry is widespread and common, especially in the tropics. Speaking of both Batesian mimicry and its corollary, Müllerian mimicry (about which see chapter 8), Lawrence Gilbert explained that the species-rich tropical rain forests "are characterized by an overwhelming variety of mimicry systems. . . . Practically every diurnally active and conspicuous arthropod participates in some form of mimicry, and the obvious cases are just a hint of what might be found using sensitive assays of auditory, visual, and chemical signals being sent, received, and imitated in the system."

The mimicry of the toxic monarch (*Danaus plexippus,* family Nymphalidae) by the viceroy (*Limenitis archippus,* family Nymphalidae) is one of the most famous examples of protective mimicry, known—at least in North America—by many people and often taught to children in school. Although the monarch and viceroy butterflies are not closely related, they look so much alike that they are hard to tell apart unless you examine them closely. The viceroy, having deviated to become a mimic, looks nothing like its closest North American relatives, almost all of which have blackish wings crossed by broad white disruptive bands. The one exception is a form of the red-spotted purple that mimics a toxic swallowtail butterfly.

The viceroy was long considered a classic example of a Batesian mimic, protected from birds only by its resemblance to the monarch. But it is not that simple. In 1958, Jane Van Zandt Brower reported that viceroys are sometimes rejected by birds, although not nearly as often as the apparently more unpalatable monarch. Thus, technically speaking, the viceroy could be more precisely called a weak Müllerian mimic of the monarch rather than a typical Batesian mimic. More than thirty years later, David Ritland and Lincoln Brower confirmed and extended Jane Brower's observations.

The viceroy actually mimics two different models, the monarch and the closely related queen butterfly (*Danaus gilippus*), both of which feed on milkweeds. Remarkably, the viceroy's appearance differs according to which of these models it coexists with. In northeastern North America, it is a dead ringer for the bright orange monarch with its prominent black wing veins. In central and southern Florida, where the monarch is scarce, the viceroy resembles the abundant queen, whose wings are dark reddish brown with veins that are not prominently black. But that's not all! In the Southwest, where the queen's wings are much lighter in color, the viceroy has correspondingly light-colored wings.

The yellow wings of all male tiger swallowtail butterflies (*Papilio glaucus*, family Papilionidae) are striped with black, the disruptive tigerlike pattern for which the species is named. Some females have the same color pattern, although the wings of other females are almost all black, with blue iridescence on the hind wings. The dark females are Batesian mimics of the toxic—but not lethal—pipevine swallowtail (*Battus philenor*, family Papilionidae), which is also mimicked more or less faithfully by female black swallowtails, female Diana fritillaries, both sexes of the spicebush swallowtail and the red-spotted purple, and by the day-flying males of the promethea moth. Female tiger swallowtails are mimetic only where they coexist with the pipevine swallowtail, in keeping with a tenet of Batesian mimicry, that mimic and model must occur together in both place and time. (We will see below that there are exceptions to this rule.) In other areas the females retain the usual tiger pattern. Where the toxic pipevine swallowtail is present, mimicking it probably gives more protection than does the camouflaging disruptive tiger pattern. However, it makes evolutionary sense to revert to second best, camouflage, in areas where the mimics are not shielded by the presence of the models.

Pipevine swallowtails live in southernmost Ontario and much of the United States except for the northern tier of states. In Florida and the Gulf states, they are joined by the closely related polydamus swallowtail (*Battus polydamus*), which—despite its name—has no tails on the hind wings. In the larval stage, both species feed only on plants of the birthwort family (Aristolochiaceae), such as Virginia snakeroot, wooly pipevine, and the cultivated Dutchman's pipe, all of which contain poisonous aristolochic acids. Just as monarchs sequester cardenolides, pipevine and polydamus caterpillars sequester aristolochic acids, which persist to the adult stage. As so elegantly demonstrated by Jane Brower, captive Florida scrub jays found pipevine swallowtails to be unpalatable, learned to reject them on sight, and thereafter also rejected mimics of the pipevine swallowtail.

The red-spotted purple (*Limenitis arthemis astyanax*, family Nymphalidae), another mimic of the pipevine swallowtail, occurs mainly or only where it coexists with its model. North of the model's range it is replaced by the white admiral (*Limenitis arthemis arthemis*), which, as we saw in chapter 4, has broad white disruptive bands on its wings. However, as my colleagues and I discovered, there is at least one major exception to this generalization. We found that on Michigan's Upper Peninsula (UP) virtually all *arthemis* are white-banded, while on the Lower Peninsula (LP) many are dark-winged pipevine swallowtail mimics, except for a few with white bands, which are often fragmentary or appear as no more than a trace. The UP and LP populations are widely separated by Lakes Michigan and Huron, except at the Straits of Mackinac, where they are only 8 kilometers apart, and some *Limenitis* make the crossing between peninsulas.

In small areas within a few kilometers of the straits, hybrids between the two forms are present. On the UP, mimiclike hybrids penetrate only a few kilometers north of the straits. Similarly, the mimics on the LP hybridize with white-banded strays from the UP, but the hybrids do not penetrate more than 20 kilometers to the south, where few individuals with even a trace of white banding are seen. This surprised us, because the entire LP is occupied—all the way to the straits—mostly by the mimetic form, which extends *350 kilometers north* of the pipevine swallowtail's range. No one knows why the disruptive form does not replace the mimics in such a large area where the mimicry's efficacy—at least theoretically—is not reinforced by the presence of the model. In the southern United States, where pipevine swallowtails are abundant, almost all female tiger swallowtails are mimetic, except in an area of central southern Florida where pipevine swallowtails are absent. North of the model's range, as in northern Michigan, they all retain the camouflaging tiger pattern. In the south of the central Midwest, where pipevine swallowtails are present but not abundant, both the mimetic and nonmimetic color phases are present.

In some species only one of the sexes, usually the female, is mimetic. No one knows why natural selection has endowed, for example, female but not male tiger swallowtails with the protection afforded by Batesian mimicry. A reasonable hypothesis, although not unequivocally supported by data, is that females are more likely to favor courting males with the ancestral color pattern of their species over those with the more recently evolved mimetic pattern. If so, any advantage the male might gain through mimicry might be more than offset by a decrease in his ability to court, win over, and inseminate females.

The sex-limited mimicry of the promethea moth (*Callosamia promethea,* family Saturniidae) is reversed. Males are Batesian mimics, almost impossible when in flight for the human eye to distinguish from a pipevine or any other dark-colored swallowtail butterfly. Females, however, look nothing like a pipevine swallowtail or any other butterfly. Their wings are a combination of light brown, dark brown, tan, and creamy white, similar to those of both sexes of many related saturniid moths. The adaptive value of this difference between the sexes is obvious. Males benefit from their mimicry because they, unlike most related moths, are active only during the daylight hours, mainly in the afternoon, and are consequently subject to attack by visually oriented, day-flying insectivorous birds. Females, however, fly only at night as they search for plants on which to lay their eggs and whose leaves their host-specific larval offspring will eat. In daylight unmated females are always out of sight, well hidden as they rest quietly in the foliage of a tree near the cocoon from which they emerged. In the afternoon they remain hidden and motionless as they release a sex-attractant pheromone. At this time the males are dangerously exposed to predatory birds as they wander in search of a pheromone odor trail that will lead them to a female. Couples stay together until evening, when the females fly for the first time in their lives as they leave their hiding places to disperse their eggs widely under the cover of darkness.

If, after I had taken my first entomology course, you had asked me if natural selection could turn a nocturnal moth into a convincing mimic of a day-flying wasp, I would have guessed that it would not be possible. But I would have been wrong. Most of the eight hundred clearwings (family Sesiidae) and a few species of other moth families are exceptionally good Batesian mimics of wasps or bees. The resemblance, Walter Linsenmaier explained, is produced not only by their long, slender, largely scale-free transparent (clear) wings, their often yellow-banded bodies, and their long, slender legs but also by their behavior. Like wasps and bees but unlike most moths, they are active during the day as they visit flowers for nectar, and they "sit and move much as . . . wasps [and bees] do and even have similar whirring and buzzing flight."

Clearwing caterpillars, according to Donald Borror and his coauthors, burrow in the roots, stems, woody trunks, or branches of trees or herbaceous plants. A few feed on cultivated plants. Larval raspberry crown borers (*Pennisetia marginata*), for example, burrow in the roots and canes of raspberries or blackberries. The adults are deceptive mimics of yellow jacket wasps. Peach tree borers (*Synanthedon exitiosa*) burrow under the bark of the trunk and are good mimics of spider wasps. Their larvae's tunneling is revealed by large masses of gum oozing from the exit of their burrow at the base of the trunk a foot or less above the soil.

I became well acquainted, much too well acquainted, with the squash vine borers (*Melittia cucurbitae*) that attacked the pattypan squash plants in my garden. The larvae tunnel in the vines but when fully grown burrow into the soil, where they spin a cocoon and spend the winter as a pupa. One summer, after they had not been molested by them for years, my squash plants suffered rather insignificant damage from just a few squash vine borers. The following spring I saw several adults emerge from the soil—just in time to glue masses of eggs all over the developing vines. After the eggs hatched that summer, the caterpillars were so numerous that they killed all my plants. But later that summer I was at least partly compensated for the loss by watching these beautiful, wasp-like moths as they laid their eggs. "The front wings," wrote Robert L. Metcalf and Robert A. Metcalf, "are covered with metallic-shining, olive-brown scales, but the hind wings are transparent. The abdomen is ringed with red, black, and copper. . . . The moth flies swiftly and noisily about the plants during the daytime, seeming more like a wasp than a moth."

Flies are more likely candidates than moths for becoming Batesian mimics of wasps or bees. Many are about the right size and shape, have elongate, scale-free (clear), membranous wings, and are day fliers that, like their models, frequently visit flowers. Members of at least eight families of flies mimic bees or wasps. Among the most notable are the insect-eating robber flies (family Asilidae), some of which are hairy and look like bumblebees. Some of the flower-visiting thick-headed flies (family Conopidae) are convincing wasp mimics. But the family Syrphidae, the hoverflies or flower flies, includes more Batesian mimics of wasps or bees than any other North American family of flies. Quite a few are what I refer to as high-fidelity mimics because they imitate even some of the finer details of the physical appearance and behavior of their models. Most low-fidelity syrphid mimics just look like flies, resembling wasps or bees only because their dark abdomens are banded with yellow. Unfortunately, it is photographs of these more common and more often seen flies that usually illustrate popular articles on mimicry. The high-fidelity yellow jacket mimic I describe next imitates both distinctive anatomical and behavioral characteristics of its model. It is seldom noticed and recognized as a fly, probably because it is mistaken for a real wasp.

So convincing is its mimicry of a yellow jacket that this large North American hoverfly (*Spilomyia hamifera*), which has no common name, deceives even experienced entomologists. I have found this—and other high-fidelity mimetic hoverflies—among the yet-to-be-identified wasps in museum insect collections. Except for their superficial but very close resemblance, these two unrelated insects have little in common. The wasp has two pairs of wings, but *Spilomyia,* like all flies, has only one pair. Yellow jackets are social, raise their young on insects in communal nests, and swarm out en masse to sting intruders. In contrast, *Spilomyia* is solitary and cannot sting, and its maggots live in and feed on the debris in wet-rot cavities in trees. However, like yellow jackets, the adults frequently drink nectar from flowers.

This fly not only has a deceptively wasplike color pattern but also, as I explained in an article in *Evolutionary Biology,* mimics—like quite a few other hoverflies—other, more detailed but salient features of its model. Like its wasp-waisted model, *Spilomyia* has a noticeably narrow waist. Like all but the mosquitoes and other, more primitive flies, it has short, three-segmented antennae that are barely visible to the naked eye, but it mimics the wasp's long, many-segmented, black and conspicuously mobile antennae by waving the black anterior segments of its long front legs in front of its head. As it sits on a blossom, the yellow jacket holds its lightly tinted, folded wings out to the sides. When thus folded lengthwise in several layers, the wings look like dark brown bands. The fly too holds its wings out to the sides. Although it can't fold them, it simulates the appearance of the wasp's folded wings with a dark brown band along the leading edge of its otherwise transparent wings. When sitting on a flower, the aposematic wasp makes itself more conspicuous by rocking from side to side. *Spilomyia* does not rock but mimics this motion by wagging its wings. Finally, if grasped by the fingers or, presumably, the beak of a bird, the fly makes a loud sound that, according to A. T. Gaul, has acoustical properties—as seen in a sonogram—almost identical to those of the squawk of a disturbed wasp.

A. Rashed and his coauthors reported that the warning squawks made by three wasp- and two honeybee-mimicking hoverflies are *all* much higher in pitch than the squawks made by the corresponding models. On this basis, they proposed that the flies are probably not acoustic mimics of wasps or bees. There is, however, another plausible—and I think more likely—interpretation of their finding. It may be that the flies' consistently higher-pitched sounds really are mimetic but are among the many examples of what animal behaviorists, such as Niko Tinbergen and Konrad Lorenz, have called supernormal stimuli. Tinbergen described a famous example in *Animal Behavior:* "Presented with two painted wooden eggs, one of normal size and the other 20 times as large, the brooding [herring] gull inspects them closely..., then tries to incubate the giant, although [the egg is too large to straddle and the gull] keeps falling off." Cartoonists create supernormal caricatures of politicians by exaggerating some of their particularly noticeable characteristics—a big nose, big ears, or an unusual hairstyle. People have said that such a caricature looks more like the person than the person him- or herself. Similarly, the mimics' "high soprano shrieks" may be supernormal stimuli even more startling to an attacking bird than the model's lower "alto squawks."

A number of insects with long antennae have models whose antennae are short, and, as Hugh Cott wrote, "it is instructive to notice the ways in which the reduction, or the appearance of reduction, in length have been achieved." For example, a Brazilian katydid mimics a wasp that has much shorter antennae. "The basal third of the [mimic's] antennae is thickened, strongly ringed, and banded... with yellow [at its tip], while the remaining two-thirds suddenly thin away to the usual hair-like fineness typical of [katydid] antennae, so that at a little distance the antennae appear to be short and yellow-tipped like those of the wasp—the basal segments alone doing duty for the entire organ of the model."

Wasp-mimicking flies have the opposite problem. Nonmimetic or low-fidelity mimetic hoverflies have the short, inconspicuous, three-segmented antennae typical of houseflies, bluebottle flies, and the other "evolutionarily advanced" flies. But as we have seen, some high-fidelity wasp-mimicking hoverflies fake the long antennae of their models by waving their front legs in front of the head. Yet even high-fidelity mimics of bumblebees or the honeybee do not do this. There is no need. Their models' antennae are long but inconspicuous because they are bent downward rather than extended forward.

Like *Spilomyia hamifera* (see figure 10), other species of its genus and some of the related genus *Temnostoma* use their front legs to simulate the long antennae of their wasp models. But, as I described in the entomological journal *Psyche,* wasp-mimicking hoverflies have at least three ways of actually lengthening their antennae without adding additional segments, which indicates that natural selection has facilitated the independent evolution of several different ways of solving the same problem. *Sphecomyia vittata*—the genus name means "wasplike fly" in Greek—has two greatly lengthened basal segments, while the terminal segment remains inconspicuous, short, and bulblike, like that of a housefly or a nonmimetic hoverfly. Another high-fidelity wasp mimic, *Ceriana signifera,* has done it differently; all three segments are greatly lengthened. *Tenthredomyia abbreviata* (the genus name is another way of saying "wasplike fly" in Greek) has evolved a third and very different way of faking long antennae. The three segments are only moderately lengthened, but the antennae seem to be much longer because they sprout from a long, thin tube that protrudes forward from the front of the head.

The yellow, black, and sometimes red colors of the hairy bumblebees are an unambiguous warning of the painful, venomous sting they use to defend themselves. They are faithfully mimicked by several species of hoverflies; the most often seen are two species of the genus *Mallota* (which have no common names). Chris Maier and I found that on sunny mornings in spring these flies often mate as they take nectar from blossoms, mainly on some of our native dogwoods and viburnums, at the edges of woodlands. In the heat of the day, they retreat to the cool shade of the woods, where the males lurk near moist rot-cavities in trees, waiting for another chance to mate by intercepting and inseminating females that come to lay eggs in them. In flight, *Mallotas* make a bumblebee-like sound, and if grasped by the fingers or trapped in the folds of an insect net, they shriek like a distressed bumblebee. Good evidence, as you will see below, shows that birds and other insectivores that have learned to reject bumblebees also reject *Mallotas* and other bumblebee mimics.

On two occasions I was fooled by one of the most convincing but improbable of the bumblebee mimics. On an early spring day I netted what I was certain was a bumblebee, but I was amazed to find nothing but a big carrion beetle (probably genus *Necrophila*, family Silphidae) in my net. It didn't look the least bit like a bumblebee. In spring, queen bumblebees—the only members of their species to survive the winter—can be seen cruising in a random pattern a few inches above the ground, searching for an abandoned animal burrow or other hole in which to found a colony. Carrion beetles behave much the same way, flying in a random pattern as they search for the lifeless carcass of an animal. Although a carrion beetle in the hand does not look the least bit like a bumblebee, it certainly does when flying with its wing covers (front wings) held above its back. The plainly visible undersides of the wing covers are an iridescent bright yellow of about the same hue as a bumblebee's yellow fur. But it is mainly the cruising beetle's unusual behavior, virtually the same as the bumblebee's equally unusual behavior, that creates the deception.

The male promethea moth looks quite a bit like a pipevine swallowtail, but the way in which it flies—like a butterfly—is also an essential part of its deception. They have deceived Jim Sternburg and me several times. My most memorable experience with one of these mimics happened while I was waiting for my wife as I sat in my car in my driveway. I "knew for sure" that I was watching a black-colored swallowtail flitting over my lawn but was astonished when it revealed itself to be a male promethea moth by trying to beat its way through a copper screen that prevented it from entering a porch on which I had placed a cage of virgin promethea females, which were then doing their best to entice males by releasing a sex-attractant pheromone.

Almost fifty years ago, Lincoln and Jane Brower proved that captive common American toads can learn not to eat bumblebees and will thereafter also reject harmless bumblebee-mimicking flies. Their experimental protocol began by determining if their naive toads were hungry by offering them a palatable dragonfly. Hungry toads, those that ate a dragonfly, were then offered a palatable bumblebee mimic, a robber fly. Those that ate the mimic—most of them—were then tempted with a real bumblebee. When they snapped up the bee, they were stung on the tongue, spit out the bee, and showed obvious signs of distress, violently wagging the tongue, ducking the head, and puffing up the body. Subsequently, these "educated" toads rejected all but one of thirty mimics presented to them, while control toads that had had no experience with bumblebees ate thirty-eight of fifty-one mimics offered to them.

The hoverfly known as the drone fly (*Eristalis tenax*) is easily mistaken for a worker honeybee, although some think that it looks more like a drone, the stingless male honeybee. There is surprisingly little evidence to support the obvious theory, that the drone fly is a Batesian mimic whose close resemblance to the stinging honeybee must to some degree protect it from birds. In 1896, C. Lloyd Morgan reported his observations of a group of young turkey chicks, to which he "first gave hive [honey] bees, which were seized, but soon let alone, and then the droneflies . . . , which so closely mimic the hive bee. They were left untouched. Their resemblance to the bees was protective. Later I gave droneflies again, and induced a chick to seize one. . . . He ran off with it chased by others. It was taken from him and swallowed. The other droneflies were then left untouched."

Systematic experiments done with other hoverflies and honeybees by the Browers almost seventy years later demonstrated that some close native North American relatives (*Eristalis vinetorum* and *Eristalis agrorum*) of the drone fly, a native of Eurasia, are protected from toads to some degree by their *generalized* resemblance to the Eurasian honeybee. It would have been more appropriate to test the responses of toads to actual drone flies, which have a *specific* resemblance to the honeybee. After all, the insects the Browers used in their experiments evolved in the Americas long before the arrival of Europeans, who brought honeybees along with them and also unintentionally introduced the drone fly. Obviously, the honeybee–drone fly mimicry system did not evolve in North America but in Eurasia, the original home of both of these insects.

The fact that drone flies have fooled humans for millennia is testimony to the near-perfection of their mimicry of the honeybee. "Because of the mimicry existing between the drone-fly and the honey-bee," E. Laurence Atkins Jr. explained, "many erroneous superstitions and myths have come to us down through the ages. For nearly three thousand years a superstition has been prevalent in the minds of the masses, as well as in the writings of the learned, to the effect that besides the usual production of honey-bees in hives, they originate by spontaneous generation from the carcasses of dead animals, and principally from those of oxen."

This bizarre myth arose because drone flies lay their eggs in putrid matter such as sewage or rotting animal carcasses. The larvae are known as rat-tailed maggots because they have a long snorkel that looks like a tail and extends up to the air above the putrescent, oxygen-deficient liquid in which they live. The many adult drone flies that swarm around a carcass are easily mistaken for bees unless closely observed. Aristotle, Atkins said, was the only learned man among the ancients who did not mention this myth in his writings about bees. He knew that four-winged insects (bees) have the sting in the tail and two-winged ones (flies) have the "sting" in the front of the head.

There were many imaginative recipes for obtaining honeybees from dead carcasses. Atkins quoted an especially elaborate one:

> Build a house ten cubits high, with all the sides of equal dimensions, with one door and four windows, one on each side; put an ox into it, thirty months old, very fat and fleshy and chosen in the spring when the sun is in the sign of the bull; let a number of young men kill him by beating violently with clubs, so as to mangle both flesh and bones but taking care not to shed any blood; let all of the orifices, mouth, eyes, nose, etc., be stopped up with clean and fine linen, impregnated with pitch—precautions to prevent the ox's vitality from escaping so that it may be conserved for the generation of the swarm of bees; let a quantity of thyme be strewed under the reclining animal, and then let the windows and doors be closed and covered with a thick coating of clay, to prevent the access of air or wind. Three weeks later let the house be opened and let the light and fresh air get access to it, except from the side from which the wind blows strongest. After eleven days you will find the house full of bees, hanging together in clusters, and nothing left of the ox but horns, bones, and hair.

This myth of the *bugonia,* which means "ox progeny" in Greek, is stated as fact in the Old Testament (Judges 14:5–9). In the vineyards of Tinmah, Samson killed a lion that "roared against him." When he passed that way some days later, "he turned aside to see the carcass of the lion: and, behold, there was a swarm of bees in the body of the lion, and honey. And he took it into his hands, and went on, eating as he went; and he came to his father and mother, and gave unto them, and they did eat." I wonder, do those who accept the words of the Bible as the literal truth ever search for honeycombs in rotting carcasses?

As we saw in chapter 8, Lincoln Brower showed that captive blue jays were made violently ill by eating a toxic monarch butterfly and thereafter refused to so much as touch one. Some retched at just the sight of one. Jane Brower had already shown that most Florida scrub jays that had been made ill by eating a monarch refused to touch viceroy butterflies, and not one ate a single mimic. Nevertheless, the ultimate test of the efficacy of Batesian mimicry is, of course, a controlled experiment done in a natural habitat with free-ranging mimics exposed to free-ranging wild birds.

Observing a statistically significant number of interactions between free-ranging wild predators, mimics, and nonmimics is almost certainly impossible. As I pointed out in a review in *Evolutionary Biology* of the attempts to measure mimetic advantage in nature, the best method would be to mark and recapture wild mimics and nonmimics, ascribing differences in recapture rates to differences in predation rates. Except with a few sedentary species, such as tropical butterflies that return to the same sleeping roost each night, recapturing marked individuals is very difficult. For example, Mike Jeffords, Jim Sternburg, and I managed to recapture with hand nets only a little more than 3 percent of the dozens of butterflies of the pipevine swallowtail Batesian complex that we had marked and released.

The ingenious promethea release-and-recapture system devised by Lincoln Brower and his coworkers nicely circumvents this difficulty. As we saw above, day-flying male promethea moths are convincing mimics, in both appearance and behavior, of the pipevine swallowtail. Unlike any of the butterflies, they are easily recaptured in early afternoon in traps baited with females releasing the sex-attractant pheromone. Because its wings are almost all black and unpatterned, a promethea male is, so to speak, a tabula rasa, a blank page, that can be painted with different colors to resemble various species of butterflies.

Over a period of years, Lincoln Brower and his colleagues did a series of experiments in Trinidad, releasing and recapturing promethea males painted to resemble a colorful toxic butterfly that is common on Trinidad and mimicked by some nontoxic butterflies. Their presumably nonaposematic controls were promethea males with black paint brushed on their black wings. This did not change their appearance, but it was a control for any possible effect on the moth of painting the wings. The Brower group ultimately concluded that their experiments had not constituted a convincing demonstration of mimetic advantage, because the artificial mimics and the presumed controls were usually recaptured in equal numbers, which indicated that they were equally likely to be captured by a predator. L. M. Cook, Brower, and John Alcock said, "Taking all the evidence over the four years there is *no significant advantage to either mimic or control moths* . . . and perhaps it should be concluded that under wild conditions no clear selective differential can be demonstrated *with the promethea moth mimicry system*" (emphasis mine).

Jim Sternburg and I reinterpreted the results of the Brower group's experiments, making what we think is a persuasive case that they did not reveal a mimetic advantage because their presumed controls, black-painted promethea males, were actually mimics of three toxic Trinidadian butterflies related and similar in appearance to the pipevine swallowtail. One of them, the polydamus swallowtail, also occurs in the southern United States, where it coexists with promethea. Consequently, the Brower group had compared two different mimics, each apparently protected to approximately the same degree by its resemblance to a different toxic butterfly. In their first publication on the promethea system, Brower and his colleagues had themselves pointed out that the male promethea may be a mimic of the pipevine swallowtail.

Adopting the Brower group's promethea release-and-recapture system, the graduate student Michael Jeffords, Sternburg, and I successfully demonstrated mimetic advantage in the pipevine swallowtail complex in central Illinois. Equal numbers of moths painted as *caricatures* of the nonmimetic and palatable yellow and black form of the tiger swallowtail and others weighted with approximately equal amounts of black paint, which did not alter their resemblance to the unpalatable pipevine swallowtail, were released in a woodland, in the center of a mile-wide circle of seven traps baited with female prometheas. The recaptured moths, more than 40 percent of the 436 released, had run the gauntlet of more that sixty species of nesting birds—all of which capture at least some flying insects—for, at the very least, half a mile, the radius of the circle of traps.

Our results leave no doubt that the moths resembling pipevine swallowtails were much more likely to survive than those resembling non-mimetic tiger swallowtails. We recaptured close to 30 percent of those that resembled the black pipevine swallowtail but less that 13 percent of those that resembled the yellow-patterned tiger swallowtail, a ratio of well over two to one. Furthermore, all of the recaptured yellow-painted but only 30 percent of the recaptured black-painted moths had wing injuries attributable to attacks by birds.

It can be argued, however, that we had not really deceived the birds, that the yellow-painted moths were attacked more often simply because they were more conspicuous than the black ones, not because they resembled tiger swallowtails. But we showed in another experiment that this was not so by releasing black-painted and yellow-painted moths, as well as orange-painted ones that resembled the unpalatable monarch. If the birds had not been deceived, the yellow- and orange-painted moths, at least to our eyes about equally conspicuous and equally unrealistic caricatures, should have suffered similar attack rates and been recaptured in about equal numbers. After all, birds, just as we do, rely mostly on their sense of vision. Yet the orange-painted moths, which resemble the toxic monarch, were recaptured much more frequently than the yellow-painted moths, which resemble the palatable tiger swallowtail—and in nearly the same numbers as the black-painted moths, which resemble the toxic pipevine swallowtail.

One of the rules of Batesian mimicry theory has long been that models and mimics must occur together at the same time and that, ideally, models should outnumber mimics. This rule seems reasonable until you consider the fact that it involves a hidden assumption: that predators, mostly birds, have short memories and soon forget an unpleasant encounter with a model. There are, however, reports of captive birds rejecting wasps, bumblebees, and their mimics for more than a year after last being stung by one.

The birds' long memories probably explain the surprising fact that high- but not low-fidelity mimics of wasps and bumblebees are, with few exceptions, present only in the spring, when their models are scarce, but not in the summer, when their models are most numerous. (Yellow jackets and bumblebees are scarce in the spring because the only ones that survive the winter are a few mated queens, which will found new colonies that will produce hundreds of workers by midsummer.) Several colleagues and I made systematic observations of the seasonal occurrence of high-fidelity bumblebee and wasp mimics and their models in three geographically different areas: a hardwood forest in central Illinois; a sand area with scrubby oaks in western Illinois; and a forest of mixed hardwoods and conifers in northern Michigan. We found thirty-one species of high-fidelity mimics (not all represented in all areas) that, with just a few exceptions, occurred only in spring in all areas, when their models were the least numerous. A few were present in late summer and a very few others (which I will come to soon) in midsummer.

How can we explain this rule-breaking finding? There may be several possible answers. For example, in the spring, flowers that produce the nectar required by the mimics are abundant, but nectar-producing flowers are also abundant in summer. There are several possible explanations, but I believe the most likely one is that high-fidelity mimics are absent during most of the summer mainly because of the threat from newly fledged young birds that are then foraging on their own for the first time but have not yet learned to avoid stinging wasps and bees. This learning is further facilitated by the absence of the harmless mimics because they do not dilute the population of stinging models during this crucial period (confirming published data and extensive unpublished records of the reproductive cycle of insectivorous birds were available for all of the research areas, for example, hundreds of field studies by students at the University of Michigan Biological Station).

We did find a few high-fidelity mimetic species in very late summer, some of which were present only then, but by that time the young birds had most likely already had painful encounters with stinging models. The few high-fidelity mimics we found in midsummer were thick-headed flies (family Conopidae), which, in the larval stage, are internal parasites of adult bumblebees or wasps. Female conopid flies tackle a flying bumblebee or wasp in midair and glue an egg to its body. After hatching, the parasitic larva burrows into its host's body. Adult conopids, excellent high-fidelity mimics of wasps, are present in midsummer because hosts for their large larvae are then much more numerous than in spring. The threat to the mimetic conopids from naive young birds is probably more than compensated for by the greater availability of hosts for their offspring.

Many nineteenth- and early-twentieth-century biologists thought that the Batesian mimicry hypothesis was false. In 1940, Hugh Cott took these naysayers to task: "The theory of mimicry has frequently been criticized as though it represented an attempt to explain an isolated, peculiar class of phenomena—championed and made much of by imaginative or over-enthusiastic arm-chair naturalists who take delight in discovering 'mimetic' resemblances in their cabinet specimens." As we saw in previous chapters, warning coloration and other aposematic signals are, to the contrary, not isolated but widespread phenomena in the animal kingdom (less so in the plant kingdom). Furthermore, it is not "arm-chair naturalists" who are conducting research on mimicry. Henry Bates and Alfred Russel Wallace were the epitome of the field naturalist. Cott, E. B. Poulton, and others were mainly field naturalists, as are Lincoln Brower, Eberhard Curio, and the author of this book. On the other hand, laboratory studies by several of these scientists, including many by Brower and his colleagues, have established that several birds, a lizard, a toad, and a few other vertebrates and even a praying mantis can learn not to eat unpalatable insects and will thereafter reject palatable mimics of these unpalatable models.

In 2003, Gabriella Gamberale-Stille and Tim Guilford considered the handicap hypothesis, proposed by A. Grafen as an alternative to the usual understanding of aposematism. Assuming that the conspicuousness of the prey's warning signal informs the predator of nothing more than that this potential meal is not worth pursuing for some unspecified reason, they proposed that "predators should associate and learn about the conspicuousness of the coloration, and not the actual colour pattern." Gamberale-Stille and Guilford tested their hypothesis by offering domestic chicks palatable crumbs and unpalatable ones sprayed with bitter quinine hydrochloride. Assuming that two contrasting colors are more conspicuous than a single color, they offered single chicks unpalatable crumbs in small containers with contrasting colors or with a single color. However, the chicks were less likely to learn to recognize and heed the presumably more conspicuous pattern, a result that, if their assumption that contrasting colors are more conspicuous to chicks is valid, refutes the handicap hypothesis.

# EPILOGUE

The ways in which insects protect themselves against insectivores and the ways in which insectivores get around these defenses are cogent examples of the reciprocal evolution of interacting groups of organisms. Not only that, they give us a look at the tactics and complexities of predation, one of the most important limits on the increase of populations and a central and necessary function in all ecosystems. Predation, parasitism, and disease are unique in being ecologically responsive, because they are, unlike climatic factors such as droughts or storms, self-limiting population regulators. As predator populations increase, for example, prey populations decrease. This shortage of the predators' food ultimately causes a steep decline in their populations, allowing prey populations, now facing less predation pressure, to rebound. Then, as their prey become more abundant, predators again become more numerous and eventually cause another decrease in the abundance of their prey. We often hear mention of the balance of nature. But as this example shows, the balance is a seesawing dynamic equilibrium.

An ecosystem with too few insects would be in a precarious state of disarray because they—the most numerous animals as both individuals and species in almost all terrestrial ecosystems—have many important, sometimes indispensable duties. As we have seen, insects are usually, by both weight and count, the most abundant and ravenous of the plant feeders. They are an essential link in food chains, making the sugars synthesized only by the photosynthetic green plants available to other insects, birds, and other animals that do not feed on green plants but do feed on insects that eat green plants.

Insects are among the most, if not *the* most, effective of the biological agents that limit the increase of plant populations. (Conversely, they also help plants by pollinating them and dispersing their seeds.) They help to keep other insect populations in check by preying on and parasitizing them. (It has often been said that insects are their own worst enemies.) They also aid in keeping bird and mammal populations within bounds by parasitizing them, sucking their blood, and transmitting their diseases. And last, but by no means least, some insects are indispensable members of an ecosystem's sanitation corps, recycling—returning to the soil—dung and dead plants and animals.

In agriculture we find many situations demonstrating that predaceous insects can greatly reduce populations of destructive plant-feeding insects, which often result from the foreseeable but too often unforeseen effects of applying insecticides that kill almost all kinds of insects to control a crop pest, often an invader from abroad: the Japanese beetle, the European corn borer, the Mediterranean fruit fly, and many others. For example, before DDT, the first of the "miracle insecticides," was used in 1946 to control the codling moth, the infamous worm in the apple, red-banded leaf roller caterpillars and European red mites had never been more than occasional minor pests in apple orchards. After apple orchards were sprayed with DDT, populations of both pests blazed out of control, and they became highly destructive. As you have no doubt realized, DDT did not kill the caterpillars or the mites, but it did kill the predaceous and parasitic insects and mites that are their natural enemies. This required the addition to the spray schedule of more insecticides and more frequent applications of them.

"Perhaps the most amazing chronicle involving pesticide-induced upsets," said Paul DeBach, "has occurred with cotton, which has been the recipient of nearly half of the agriculture insecticides used in the United States, and is similarly heavily treated in many [other] countries." In Mexico, for instance, as more and more different kinds of insecticides were added to the spray schedule to control "upsets," even more upsets occurred, necessitating more and more insecticide applications and the addition of more kinds of insecticides. In some areas cotton culture was abandoned because the investment in insecticides had become too expensive.

The preceding examples demonstrate that predator-prey interactions are a central aspect of ecosystems. Understanding the interactions between insect eaters and their prey is clearly essential to understanding terrestrial and freshwater ecosystems, because of the insects' significant, often critical, roles in them. As is to be expected, over hundreds of millions of years there has been an escalating arms race between predators and prey, as exemplified in the ten chapters of this book. But how did the many weapons of the predators and the counterdefenses of the prey evolve? Batesian mimicry is arguably the ultimate of the counterdefenses that have evolved so far. After all, it could not have evolved until after insects and other animals had first evolved defenses such as venoms and other toxins and ways of advertising—aposematism—these defenses to birds and other predators.

Mimicry has been of special interest to evolutionary biologists. Many have speculated on how Batesian mimicry evolves, among them Tim Guilford and others cited by Graeme Ruxton and his coauthors. From the first, the consensus has been that mimicry evolves in the usual Darwinian fashion, by the gradual accretion, via natural selection, of small changes that progressively refine the mimic's resemblance to its model. The evolutionary path of mimicry is unusually—perhaps uniquely—perspicuous, because many of the characteristics that natural selection is likely to favor are predictable: just about anything that improves the mimic's resemblance to its model's appearance or behavior. In his preface to *Mimicry and the Evolutionary Process,* Lincoln Brower reasserted the opinion of the distinguished evolutionary biologist R. A. Fisher that protective mimicry provides the most clear-cut post-Darwinian example of natural selection.

However, Richard Goldschmidt disagreed. In a 1963 article, Brower and his coauthors succinctly summarized Goldschmidt's argument "that mimicry can not arise by the accumulation of small variations because the initial [minor] changes would" offer no protection against "predators which had experienced the potential models." Accordingly, Goldschmidt proposed that mimicry could evolve only in big jumps that produce a near-perfect mimic in one step. Years later, Stephen J. Gould and Niles Eldredge resurrected Goldschmidt's notion that evolution proceeds by leaps. Their theory of punctuated equilibrium states that over a long period of time organisms acquire, via natural selection, some minor attributes that help to adapt them to their environment. But only at the end of this long period does a sudden brief burst of major changes transform them into a separate species that is "reproductively isolated," its members capable of mating and reproducing only with members of their own species. Gould and his various coauthors have argued that these postulated leaps are not brought about by natural selection and are, therefore, not adaptations to the environment. They are the result of some unknown force, a deus ex machina that always was and still remains a mystery. This theory was inspired by Gould and Eldredge's interpretation of the fossil record, in which "they saw ... long geological periods during which fossil species change but little, followed by brief periods characterized by major changes." "Others," Monroe Strickberger pointed out, "argue that because the fossil record has gaps in fossilization, punctuated equilibria are only apparent, not real."

Brower and his coauthors refuted Goldschmidt—as well as Gould's yet to be enunciated theory of punctuated equilibrium—by citing laboratory and field observations and experiments done by him and his coworkers as well as many others. Referring to Goldschmidt's contention that minor resemblances to a toxic organism would give no protection to an incipient mimic—which is also implied by Gould's hypothesis—they said, "The contrary may now be stated as an experimentally demonstrated fact, namely that even a remote resemblance between ... butterflies can be advantageous." Experiments by other scientists, many reviewed by Ruxton and his coauthors, similarly establish that even incipient mimics receive some protection, although much less than high-fidelity mimics. For example, Jane Van Zandt Brower found that scrub jays that had learned to reject monarchs also rejected not only the monarch-mimicking viceroy but also a somewhat different form of the viceroy—the very same species—that mimics a dark-colored relative of the monarch, the queen, which looks different from the monarch. The jays had not been exposed to queens and, consequently, had not learned to reject them.

In 1963, Ernst Mayr, the twentieth century's outstanding practitioner of systematics (the science of the evolution and classification of organisms) proclaimed that the formation of new species is always allopatric (occurring in separate places). That is, only when two populations of a species cannot commingle because they are separated by a barrier—a wide river, a mountain range—can they become reproductively isolated, unable to interbreed, the hallmark of a species.

Charles Darwin had suggested that two sympatric populations (those commingling in the same habitat) could become reproductively isolated if they were ecologically separated. This concept fell into disrepute, but as reported by Stewart Berlocher and Jeffry Feder, it was brilliantly revived by Guy Bush, who used the native American fruit fly as an example. This insect, the apple maggot, which originally fed only on fruits of the native hawthorn, eventually also fed on fruits of the related but nonnative apple. Today the apple- and hawthorn-feeding host races mingle in the same habitat. The adult flies usually orient to the kind of plant they fed on in the larval stage. Females stab their eggs into the fruits, on which males lie in wait to waylay and inseminate the arriving females. Consequently, the two host races are well on their way to becoming reproductively isolated species. Experiments done in nature show that only about 5 percent of matings are between apple- and hawthorn-feeding individuals.

It seems clear that Goldschmidt was wrong. That mimickry evolves in the usual Darwinian fashion, by the accretion of small changes.

# SELECTED REFERENCES

## PROLOGUE

Edmunds, M. 1974. *Defence in Animals.* New York: Longman Group.

Krebs, J. R., and N. B. Davies. 1993. *An Introduction to Behavioural Ecology.* Oxford: Blackwell Scientific.

## 1. INSECTS IN THE WEB OF LIFE

Bates, H. W. 1862. Contributions to an insect fauna of the Amazon Valley, Lepidoptera: Heliconidae. *Transactions of the Linnaean Society, Zoology* 23: 495–566.

Berenbaum, M. R. 1995. *Bugs in the System: Insects and Their Impact on Human Affairs.* Reading, MA: Addison-Wesley.

Buchmann, S. L., and G. P. Nabhan. 1996. *The Forgotten Pollinators.* Washington, DC: Shearwater Books.

Hocking, B. 1968. *Six-Legged Science.* Cambridge, MA: Schenkman.

Marshall, S. A. 2006. *Insects: Their Natural History and Diversity.* Buffalo, NY: Firefly Books.

Odum, E. P. 1971. *Fundamentals of Ecology.* 3rd ed. Philadelphia: Saunders.

Price, P. W. 1997. *Insect Ecology.* 3rd ed. New York: John Wiley and Sons.

Williams, C. M. 1958. Hormonal regulation of insect metamorphosis. In *A Symposium on the Chemical Basis of Development,* ed. W. D. McElroy and G. Glass. Baltimore: Johns Hopkins University Press.

## 2. THE EATERS OF INSECTS

Annandale, N. 1900. Observations on the habits and natural surroundings of insects made during the Skeat 'Expedition' to the Malay Peninsula, 1899–1900. *Proceedings of the Zoological Society of London* 1900: n.p.

Askew, R.R. 1971. *Parasitic Insects*. New York: American Elsevier.

Balduf, W.V. 1939. *The Bionomics of Entomophagous Insects*. St. Louis: John F. Swift.

Bastin, H. 1913. *Insects: Their Life-Histories and Habits*. New York: Frederick A. Stokes.

Boswall, J. 1977. Tool-using by birds and related behaviour. *Avicultural Magazine* 83: 88–97, 146–59, 220–28.

———. 1983. Tool-using and related behaviour in birds: More notes. *Avicultural Magazine* 89: 94–108.

Brandt, H. 1951. *Arizona and Its Bird Life*. Cleveland: privately printed by the Bird Research Foundation.

Bristowe, W.S. 1976. *The World of Spiders*. London: Collins.

Burt, W.H. 1957. *Mammals of the Great Lakes Region*. Ann Arbor: University of Michigan Press.

Clausen, C.P. 1952. Parasites and predators. In *Insects, the Yearbook of Agriculture*. Washington, DC: U.S. Department of Agriculture.

Cott, H.B. 1957. *Adaptive Coloration in Animals*. London: Methuen.

Dejean, A., P.J. Solano, J. Ayroles, B. Cobara, and J. Orivel. 2005. Arboreal ants build traps to capture prey. *Nature* 434: 973.

Eisner, T., R. Alsop, and G. Ettershank. 1964. Adhesiveness of spider silk. *Science* 146: 1058–61.

Emery, N.J., and N.S. Clayton. 2004. The mentality of crows: Convergent evolution in corvids and apes. *Science* 306: 1903–7.

Felt, E.P. 1905. *Insects Affecting Park and Woodland Trees*. Memoir 8 of the New York State Museum. Albany: New York State Education Department.

Foelix, R.F. 1982. *Biology of Spiders*. Cambridge, MA: Harvard University Press.

Forbush, E.H., and J.B. May. 1939. *Natural History of the Birds of Eastern and Central North America*. Boston: Houghton Mifflin.

Fraenkel, G.S., and F. Fallil. 1981. The spinning (stitching) behaviour of the rice leaf folder, *Cnaphalocrosis medinalis*. *Entomologia Experimentalis et Applicata* 29: 138–46.

Gause, G.F. 1934. *The Struggle for Existence*. Baltimore: Williams and Wilkins.

Gill, F.B. 1995. *Ornithology*. 2nd ed. New York: W.H. Freeman.

Goodall, J. 1963. Feeding behavior of wild chimpanzees. *Symposia of the Zoological Society of London* 10: 39–47.

Heinrich, B., and S. Collins. 1983. Caterpillar leaf damage and the game of hide-and-seek with birds. *Ecology* 64: 592–602.

Holmes, R. T., J. C. Schultz, and P. Nothnagle. 1979. Bird predation on forest insects: An exclosure experiment. *Science* 206: 462–63.

Lack, D. 1947. *Darwin's Finches*. Cambridge: Cambridge University Press.

Levy, D. L., R. S. Duncan, and C. F. Levins. 2004. Use of dung as a tool by burrowing owls. *Science* 431: 39.

MacArthur, R. H. 1958. Population ecology of some warblers of northeastern coniferous forests. *Ecology* 39: 599–619.

Marquis, R. J., and C. J. Whelan. 1994. Insectivorous birds increase growth of white oak through consumption of leaf-chewing insects. *Ecology* 75: 2007–14.

Martin, I. G. 1981. Venom of the short-tailed shrew *(Blarina brevicauda)* as an insect immobilizing agent. *Journal of Mammalogy* 62: 182–91.

Metcalf, R. L., and R. A. Metcalf. 1993. *Destructive and Useful Insects*. 5th ed. New York: McGraw-Hill.

Montgomery, S. L. 1982. Biogeography of the moth genus *Eupithecia* in Oceania and the evolution of ambush predation in Hawaiian caterpillars (Lepidoptera: Geometridae). *Entomologia Generalis* 8: 27–34.

Morse, D. H. 1968. The use of tools by brown-headed nuthatches. *Wilson Bulletin* 80: 220–24.

Nadis, S. 2006. Hard-hitting endeavour captures Ig Nobel. *Nature* 443: 616–17.

Peterson, R. T. 1963. *The Birds*. New York: Time.

Smith, S. M. 1991. *The Black-Capped Chickadee*. Ithaca, NY: Cornell Univ. Press.

Sullivan, K. A. 1984. Information exploitation by downy woodpeckers in mixed-species flocks. *Behaviour* 91: 294–311.

Tebbich, S., M. Taborsky, B. Fessl, and D. Blomqvist. 2001. Do woodpecker finches acquire tool-use by social learning? *Proceedings of the Royal Society of London B* 268: 2189–93.

Tinbergen, N. 1965. *Animal Behavior*. New York: Time-Life Books.

Van Tyne, J. 1951. A cardinal's, *Richmondena cardinalis*, choice of food for adult and for young. *Auk* 68: 110.

Waldbauer, G. P. 2009. *Fireflies, Honey, and Silk*. Berkeley: University of California Press.

Wheeler, W. M. 1930. *Demons of the Dust*. New York: W. W. Norton.

Wickler, W. 1968. *Mimicry in Plants and Animals*. Translated from the German by R. D. Martin. New York: McGraw-Hill.

3. FLEEING AND STAYING UNDER COVER

Angelon, K.A., and J.W. Petranka. 2002. Chemicals of predatory mosquitofish (*Gambusia affinis*) influence selection of oviposition site by *Culex* mosquitoes. *Journal of Chemical Ecology* 28: 797–806.

Ball, H.J. 1965. Photosensitivity in the terminal abdominal ganglion of *Periplaneta americana* (L.). *Journal of Insect Physiology* 11: 1311–15.

Bastin, H. 1913. *Insects: Their Life-Histories and Habits.* New York: Frederick A. Stokes.

Berenbaum, M.R. 1995. *Bugs in the System: Insects and Their Impact on Human Affairs.* Reading, MA: Addison-Wesley.

Bruno, M.S., and D. Kennedy. 1962. Spectral sensitivity of photoreceptor neurons in the sixth ganglion of the crayfish. *Comparative Biochemistry and Physiology* 6: 41–46.

Burt, W.H. 1957. *Mammals of the Great Lakes Region.* Ann Arbor: University of Michigan Press.

Callahan, P.S. 1965. A photoelectric-photographic analysis of flight behavior in the corn earworm, *Heliothis zea,* and other moths. *Annals of the Entomological Society of America* 58: 159–69.

Chapman, R.F. 1971. *The Insects: Structure and Function.* 2nd ed. New York: Elsevier.

Comstock, J.H. 1950. *An Introduction to Entomology.* 9th ed. Ithaca, NY: Comstock Publishing.

Conner, J., S. Camazine, D. Aneshansley, and T. Eisner. 1985. Mammalian breath: Trigger of defensive chemical response in a tenebrionid beetle (*Bolitotherus cornutus*). *Behavioral Ecology and Sociobiology* 16: 115–18.

Dickinson, M.H., and J.R.B. Lighton. 1995. Muscle efficiency and elastic storage in the flight motor of *Drosophila*. *Science* 26: 87–90.

Edmunds, M. 1974. *Defence in Animals.* New York: Longman.

Eisner, T., M. Eisner, and M. Siegler. 2005. *Secret Weapons: Defenses of Insects, Spiders, Scorpions, and Other Many-Legged Creatures.* Cambridge, MA: Harvard University Press.

Evans, H.E. 1984. *Insect Biology.* Reading, MA: Addison-Wesley.

Felt, E.P. 1905. *Insects Affecting Park and Woodland Trees.* Memoir 8 of the New York State Museum. Albany: New York State Education Department.

Fitzgerald, T.D. 1995. *The Tent Caterpillars.* Ithaca, NY: Cornell University Press.

Frisch, K. von. 1974. *Animal Architecture.* New York: Harcourt Brace Jovanovich.

Gill, F. B. 1995. *Ornithology*. 2nd ed. New York: W. H. Freeman.

Graham, S. A. 1952. *Forest Entomology*. New York: McGraw-Hill.

Griffin, D. R. 1958. *Listening in the Dark*. New Haven, CT: Yale University Press.

Hamilton, W. D. 1971. Geometry for the selfish herd. *Journal of Theoretical Biology* 31: 295–311.

Hughes, G. M., and P. J. Mill. 1974. Locomotion: Terrestrial. In *The Physiology of Insecta*, ed. M. Rockstein. 2nd ed. New York: Academic Press.

Linsenmaier, W. 1972. *Insects of the World*. Translated from the German by L. E. Chadwick. New York: McGraw-Hill.

Marshall, S. A. 2006. *Insects: Their Natural History and Diversity*. Buffalo, NY: Firefly Books.

Matthews, R. W., and J. R. Matthews. 1978. *Insect Behavior*. New York: John Wiley and Sons.

McConnell, E., and A. G. Richards. 1955. How fast can a cockroach run? *Bulletin of the Brooklyn Entomological Society* 50: 36–43.

Metcalf, R. L., and R. A. Metcalf. 1993. *Destructive and Useful Insects*. 5th ed. New York: McGraw-Hill.

Peterson, R. T. 1963. *The Birds*. New York: Time.

Roeder, K. D. 1963. *Nerve Cells and Insect Behavior*. Cambridge, MA: Harvard University Press.

Scarbrough, A. G., G. P. Waldbauer, and J. G. Sternburg. 1972. Response to cecropia cocoons of *Mus musculus* and two species of *Peromyscus*. *Oecologia* 10: 137–44.

Schmitz, O. J. 2008. Effects of predator hunting mode on grassland ecosystem function. *Science* 319: 952–54.

Tinbergen, N. 1965. *Animal Behavior*. New York: Time-Life Books.

Waage, J. K. and G. G. Montgomery. 1976. *Cryptoses choloepi*: A coprophagous moth that lives on a sloth. *Science* 193: 157–58.

Waldbauer, G. P., and J. G. Sternburg. 1982. Cocoons of *Callosamia promethea* (Saturniidae): Adaptive significance of differences in mode of attachment to the host tree. *Journal of the Lepidopterist's Society* 36: 192–99.

Waldbauer, G. P., J. G. Sternburg, W. G. George, and A. G. Scarbrough. 1970. Hairy and downy woodpecker attacks on cocoons of urban *Hyalophora cecropia* and other saturniids (Lepidoptera). *Annals of the Entomological Society of America* 63: 1366–69.

Wigglesworth, V. B. 1972. *The Principles of Insect Physiology*. 7th ed. London: Chapman and Hall.

4. HIDING IN PLAIN SIGHT

Brower, L.P., and J.V.Z. Brower. 1956. Cryptic coloration in the anthophilous moth *Rhododipsa masoni. American Naturalist* 90: 177–82.

Cott, H.B. 1957. *Adaptive Coloration in Animals.* London: Methuen.

Cuthill, I.C., M. Stevens, J. Sheppard, T. Maddocks, C.A. Párraga, and T. Troscianko. 2005. Disruptive coloration and background pattern matching. *Nature* 434: 72–74.

De Ruiter, L. 1952. Some experiments on the camouflage of stick caterpillars. *Behaviour* 4: 222–32.

Di Cesnola, A.P. 1904. Preliminary note on the protective value of colour in *Mantis religiosa. Biometrika* 3: 58–59.

Edmunds, M. 1974. *Defence in Animals.* New York: Longman.

Forbush, E.H., and C.H. Fernald. 1896. *The Gypsy Moth.* Boston: Massachusetts State Board of Agriculture.

Gerould, J.H. 1921. Blue-green caterpillars: The origin and ecology of a mutation in hemolymph color in *Colias (Eurymus) philodice. Journal of Experimental Zoology* 34: 385–416.

Grant, B., D.F. Owen, and C.A. Clarke. 1995. Decline of melanic moths. *Nature* 373: 565.

Grant, B., and L.L. Wiseman. 2002. Recent history of melanism in American peppered moths. *Journal of Heredity* 93: 86–90.

Hazel, W., S. Ante, and B. Stringfellow. 1998. The evolution of environmentally-cued pupal colour in swallowtail butterflies: Natural selection for pupation site and pupal colour. *Ecological Entomology* 23: 41–44.

Himmelman, J. 2002. *Discovering Moths.* Camden, ME: Down East Books.

Hingston, R.W.G. 1932. *A Naturalist in the Guiana Forest.* New York: Longmans, Green.

Holland, W.J. [1903] 1920. *The Moth Book.* Reprint, New York: Doubleday, Page.

Isely, F.C. 1938. Survival value of acridian protective coloration. *Ecology* 19: 370–89.

Kettlewell, H.B.D. 1959. Darwin's missing evidence. *Scientific American* 200: 48–53.

Linsenmaier, W. 1972. *Insects of the World.* Translated from the German by L.E. Chadwick. New York: McGraw-Hill.

Lutz, F.E. 1935. *Field Book of Insects.* New York: G.P. Putnam's Sons.

Marshall, S.A. 2006. *Insects: Their Natural History and Diversity.* Buffalo, NY: Firefly Books.

Moffett, M. W. 2007. Able bodies. *National Geographic,* August, 140–50.

Owen, D. F. 1980. *Camouflage and Mimicry.* Chicago: University of Chicago Press.

Sargent, T. D. 1976. *Legion of Night: The Underwing Moths.* Amherst: University of Massachusetts Press.

Silberglied, R. E., A. Aiello, and D. M. Windsor. 1980. Disruptive coloration in butterflies: Lack of support in *Anartia fatima. Science* 209: 617–19.

Waldbauer, G. P., and J. G. Sternburg. 1983. A pitfall in using painted insects in studies of protective coloration. *Ecology* 37: 1085–86.

Waldbauer, G. P., J. G. Sternburg, and A. W. Ghent. 1988. Lakes Michigan and Huron limit gene flow between the subspecies of the butterfly *Limenitis arthemis. Canadian Journal of Zoology* 66: 1790–95.

Williams, C. M. 1958. Hormonal regulation of insect metamorphosis. In *A Symposium on the Chemical Basis of Development,* ed. W. D. McElroy and G. Glass. Baltimore: Johns Hopkins University Press.

## 5. BIRD DROPPING MIMICRY AND OTHER DISGUISES

Bastin, H. 1913. *Insects: Their Life-Histories and Habits.* New York: Frederick A. Stokes.

Comstock, J. H. 1950. *An Introduction to Entomology.* 9th ed. Ithaca, NY: Comstock Publishing.

Cott, H. B. 1957. *Adaptive Coloration in Animals.* London: Methuen.

Edmunds, M. 1974. *Defence in Animals.* New York: Longman Group.

Forbes, H. O. 1885. *A Naturalist's Wanderings in the Eastern Archipelago.* London: Sampson, Low, Marston, Searle, and Rivington.

Gregory, J. W. 1896. *The Great Rift Valley.* London: J. Murray.

Himmelman, J. 2002. *Discovering Moths.* Camden, ME: Down East Books.

Hingston, R. W. G. 1932. *A Naturalist in the Guiana Forest.* New York: Longmans, Green.

Newnham, A. 1924. The detailed resemblance of an Indian lepidopterous larva to the excrement of a bird. A similar result obtained in an entirely different way by a Malayan spider. *Transactions of the Entomological Society of London* 1924: xc–xciv.

Opler, P. A., and G. O. Krizek. 1984. *Butterflies East of the Great Plains.* Baltimore: Johns Hopkins University Press.

Owen, D. F. 1980. *Camouflage and Mimicry.* Chicago: University of Chicago Press.

## 6. FLASH COLORS AND EYESPOTS

Annandale, N. 1900. Observations on the habits and natural surroundings of insects made during the "Skeat Expedition" to the Malay Peninsula, 1899–1900. *Proceedings of the Zoological Society of London* 1900: 837–69.

Blest, A. D. 1957. The function of eyespot patterns in the Lepidoptera. *Behaviour* 11: 209–55.

Bouseman, J. K., and J. G. Sternburg. 2002. *Field Guide to Silkmoths of Illinois.* Illinois Natural History Survey Manual 10. Champaign: Illinois Natural History Survey.

Cott, H. B. 1957. *Adaptive Coloration in Animals.* London: Methuen.

Curio, E. 1965. Ein Falter mit falschem Kopf [A butterfly with a false head]. *Natur und Museum* 95: 43–46.

Edmunds, M. 1974. *Defence in Animals.* New York: Longman Group.

Eisner, T. 2003. *For Love of Insects.* Cambridge, MA: Harvard University Press.

Farb, P. 1962. *The Insects.* New York: Time.

Maldonado, H. 1970. The deimatic reaction in the praying mantis *Stagmatoptera biocellata. Zeitschrift für vergleichende Physiologie* 68: 60–71.

Opler, P. A., and G. O. Krizek. 1984. *Butterflies East of the Great Plains.* Baltimore: Johns Hopkins University Press.

Owen, D. F. 1980. *Camouflage and Mimicry.* Chicago: University of Chicago Press.

Robbins, R. K. 1981. The "false head" hypothesis: Predation and wing pattern variation of Lycaenid butterflies. *American Naturalist* 118: 770–75.

Ruxton, G. D., T. N. Sherratt, and M. P. Speed. 2004. *Avoiding Attack.* Oxford: Oxford University Press.

Sargent, T. D. 1976. *Legion of Night: The Underwing Moths.* Amherst: University of Massachusetts Press.

Schlenoff, D. H. 1985. The startle responses of blue jays to *Catocala* (Lepidoptera: Noctuidae) prey models. *Animal Behaviour* 33: 1057–67.

Stevens, M., E. Hopkins, W. Hinde, A. Adcock, Y. Connolly, T. Troscianko, and I. C. Cuthill. 2007. Field experiments on the effectiveness of "eyespots" as predator deterrents. *Animal Behaviour* 74: 1215–27.

Vallin, A., S. Jakobsson, J. Lind, and C. Wiklund. 2005. Prey survival by predator intimidation: An experimental study of peacock butterfly defence against blue tits. *Proceedings of the Royal Society B* 272: 1203–7.

Vaughan, F.A. 1983. Startle responses of blue jays to visual stimuli presented during feeding. *Animal Behaviour* 31: 385–96.

Wickler, W. 1968. *Mimicry in Plants and Animals.* Translated from the German by R.D. Martin. New York: McGraw-Hill.


7. SAFETY IN NUMBERS

Ashall, C., and P.E. Ellis. 1962. *Studies on Numbers and Mortality in Field Populations of the Desert Locust.* Anti-Locust Bulletin 38. London: Anti-Locust Research Centre.

Buck, J.B., and E. Buck. 1976. Synchronous fireflies. *Scientific American* 234: 74–85.

Dinesen, I. 1937. *Out of Africa.* New York: Modern Library.

Eisner, T., J.S. Johnessee, J. Carrel, L.B. Hendry, and J. Meinwald. 1974. Defensive use by an insect of a plant resin. *Science* 184: 996–99.

Ellis, P.E. 1959. Learning and social aggregation in locust hoppers. *Animal Behaviour* 7: 91–106.

Evans, H.E. 1966. The accessory burrows of digger wasps. *Science* 152: 465–71.

———. 1966. The behavior patterns of solitary wasps. *Annual Review of Entomology* 11: 128–54.

Foster, W.A., and J.E. Treherne. 1981. Evidence for the dilution effect in the selfish herd from fish predation on a marine insect. *Nature* 293: 466–67.

Ghent, A.W. 1960. A study of the group-feeding behaviour of larvae of the jack pine sawfly, *Neodiprion pratti banksianae* Roh. *Behaviour* 16: 110–48.

Gillett, S.D. 1988. Solitarization in the desert locust, *Schistocerca gregaria* (Forskål) (Orthoptera: Acrididae). *Bulletin of Entomological Research* 78: 623–31.

Hamilton, W.D. 1971. Geometry of the selfish herd. *Journal of Theoretical Biology* 31: 295–311.

Hogue, C.L. 1972. Protective function of sound perception and gregariousness in *Hylesia* larvae (Saturniidae: Hemileucinae). *Journal of the Lepidopterists' Society* 26: 33–34.

Hudleston, J.A. 1958. Some notes on the effects of bird predators on hopper bands of the desert locust (*Schistocerca gregaria* Forskål). *Entomologist's Monthly* 94: 110–14.

Ischii, S. 1970. Aggregation of the german cockroach, *Blatella germanica* (L.). In *Control of Insect Behavior by Natural Products,* ed. D.L. Wood, R.M. Silverstein, and M. Nakajama. New York: Academic Press.

Lockwood, J.A., and R.N. Story. 1985. Bifunctional pheromone in the first instar of the southern green stink bug, *Nezara viridula* (L.) (Hemiptera: Pentatomidae): Its characterization and interaction with other stimuli. *Annals of the Entomological Society of America* 78: 474–79.

―――. 1987. Defensive secretion of the southern green stink bug (Hemiptera: Pentatomidae) as an alarm pheromone. *Annals of the Entomological Society of America* 80: 686–91.

Matthews, R.W., and J.R. Matthews. 1978. *Insect Behavior.* New York: John Wiley and Sons.

Metcalf, R.L., and R.A. Metcalf. 1993. *Destructive and Useful Insects.* 5th ed. New York: McGraw-Hill.

Michener, C.D. 1974. *The Social Behavior of the Bees.* Cambridge, MA: Harvard University Press.

Myers, J.H., and J.N. Smith. 1978. Head flicking by tent caterpillars: A defensive response to parasite sounds. *Canadian Journal of Zoology* 56: 1628–31.

Sweeney, B.W., and R.L. Vannote. 1982. Population synchrony in mayflies: A predator satiation hypothesis. *Evolution* 36: 810–21.

Tanaka, S., H. Wolda, and D.L. Denlinger. 1988. Group size affects the metabolic rate of a tropical beetle. *Physiological Entomology* 13: 239–41.

Tinbergen, N., M. Impekoven, and D. Franck. 1967. An experiment on spacing-out as a defence against predation. *Behaviour* 28: 307–21. Reprinted in N. Tinbergen, *The Animal in Its World.* Vol. 1. Cambridge, MA: Harvard University Press, 1972.

Uvarov, B. 1921. A revision of the genus *Locusta*, L. ( = *Pachystylus,* Fieb.) with a new theory as to the periodicity and migrations of locusts. *Bulletin of Entomological Research* 12: 135–63.

Vulinec, K. 1990. Collective security: Aggregation by insects as a defense. In *Insect Defenses,* ed. D.L. Evans and J.O. Schmidt. Albany: State University of New York Press.

Williams, C.B. 1958. *Insect Migration.* New York: Macmillan.

Wilson, E.O. 1971. *The Insect Societies.* Cambridge, MA: Harvard University Press.

## 8. DEFENSIVE WEAPONS AND WARNING SIGNALS

Akre, R.D., A. Greene, J.F. MacDonald, P.J. Landolt, and H.G. Davis. 1980. *The Yellowjackets of America North of Mexico.* Washington, DC: U.S. Department of Agriculture.

Aneshansley, D.J., T. Eisner, J.M. Widom, and B. Widom. 1969. Biochemistry at 100°C: Explosive secretory discharge of bombardier beetles *(Brachinus)*. *Science* 165: 61–63.

Attygalle, A.B., S.R. Smedley, J. Meinwald, and T. Eisner. 1993. Defensive secretion of two notodontid caterpillars *(Schizura unicornis, S. badia)*. *Journal of Chemical Ecology* 19: 2089–104.

Bastin, H. 1913. *Insects: Their Life-Histories and Habits.* New York: Frederick A. Stokes.

Berenbaum, M.R. 1995. *Bugs in the System: Insects and Their Impact on Human Affairs.* Reading, MA: Addison-Wesley

Berenbaum, M.R., and E. Miliczky. 1984. Mantids and milkweed bugs: Efficacy of aposematic coloration against invertebrate predators. *American Midland Naturalist* 111: 64–68.

Bishop, H. 2005. *Robbing the Bees.* New York: Free Press.

Bouseman, J.K., and J.G. Sternburg. 2002. *Field Guide to Silkmoths of Illinois.* Illinois Natural History Survey Manual 10. Champaign: Illinois Natural History Survey.

Bowers, M.D. 1993. Aposematic caterpillars: Life-styles of the warningly colored and unpalatable. In *Caterpillars,* ed. N.E. Stamp and T.M. Casey. New York: Chapman and Hall.

Boyden, T.C. 1976. Butterfly palatability and mimicry: Experiments with *Ameiva* lizards. *Evolution* 30 73–81.

Brower, L.P. 1969. Ecological chemistry. *Scientific American* 220: 4–15.

Brown, S.G., G.H. Boettner, and J.E. Yack. 2007. Clicking caterpillars: Acoustic aposematism in *Antheraea polyphemus* and other Bombicoidea. *Journal of Experimental Biology* 210: 993–1005.

Davies, H., and C.A. Butler. 2008. *Do Butterflies Bite?* New Brunswick, NJ: Rutgers University Press.

Dussourd, D.E. 1993. Foraging with finesse: Caterpillar adaptations for circumventing plant defenses. In *Caterpillars,* ed. N.E. Stamp and T.M. Casey. New York: Chapman and Hall.

Edmunds, M. 1974. *Defence in Animals.* New York: Longman Group.

Edwards, J.S. 1960. Spitting as a defensive mechanism in a predatory reduviid. *Proceedings of the XI International Congress of Entomology* 3: 259–63.

Eisner, T. 1965. Defensive spray of a phasmid insect. *Science* 148: 966–68.

Eisner, T., M. Eisner, and M. Siegler. 2005. *Secret Weapons: Defenses of Insects, Spiders, Scorpions, and Other Many-Legged Creatures.* Cambridge, MA: Harvard University Press.

Eisner, T., J. Meinwald, A. Monro, and R. Ghent. 1961. Defence mechanisms of anthropods, I. The composition and function of the spray of the whip-scorpion *Mastigoproctus giganteus* (Lucas) (Arachnida, Pedipalpida). *Journal of Insect Physiology* 6: 272–98.

Eltringham, H. 1913. On the urticating properties of *Porthesia similis*, Fuess. *Transactions of the Entomological Society of London* 1913: 423–27.

Ford, E. B. 1955. *Moths.* London: Collins.

Gamberale-Stille, G., and T. Guilford. 2003. Contrast versus colour in aposematic signals. *Animal Behaviour* 65: 1021–26.

Guilford, T. 1988. The evolution of conspicuous coloration. In *Mimicry and the Evolutionary Process,* ed. L. P. Brower, pp. 7–21. Chicago: University of Chicago Press.

Hamilton, W.D. 1964. The evolution of social behavior. *Journal of Theoretical Biology* 7: 1–52.

Hoffmeister, D. F., and C. O. Mohr. 1972. *Fieldbook of Illinois Mammals.* New York: Dover.

Hölldobler, B., and E. O. Wilson. 1990. *The Ants.* Cambridge, MA: Harvard University Press.

Linsenmaier, W. 1972. *Insects of the World.* Translated from the German by L. E. Chadwick. New York: McGraw-Hill.

Lloyd, J. E. 1975. Aggressive mimicry in *Photuris* fireflies: Signal repertoires by femmes fatales. *Science* 187: 452–53.

Marshall, S.A. 2006. *Insects, Their Natural History and Diversity.* Buffalo, NY: Firefly Books.

Michener, C.D. 1974. *The Social Behavior of the Bees.* Cambridge, MA: Harvard University Press.

Müller, F. 1879. *Ituna* and *Thyridia;* a remarkable case of mimicry in butterflies. Translated from the German by R. Meldola. *Proceedings of the Entomological Society of London* 27: xx–xxix.

Nishida, R. 2002. Sequestration of defensive substances from plants by Lepidoptera. *Annual Review of Entomology* 45: 57–92.

Ratcliffe, J. R., and M. L. Nydam. 2008. Multimodal warning signals for a multiple predator world. *Nature* 455: 96–99.

Remold, H. 1963. Scent-glands of land-bugs, their physiology and biological function. *Nature* 198: 764–66.

Rothschild, M. 1972. Secondary plant substances and warning colouration in insects. In *Insect/Plant Relationships,* ed. H. F. van Emden. Symposium 6 of the Royal Entomological Society of London.

———. 1997. Discovering details. *Wing* 20: 13–18.

Rothschild, M., J. von Euw, and T. Reichstein. 1970. Cardiac glycosides in the oleander aphid. *Journal of Insect Physiology* 16: 1141–45.

———. 1973. Cardiac glycosides in a scale insect *(Aspidiotus)*, a ladybird *(Coccinella)* and a lacewing *(Chrysopa)*. *Journal of Entomology (A)* 48: 89–90.

Rothschild, M., and P.T. Haskell. 1966. Stridulation of the garden tiger moth, *Arctia caja* L, audible to the human ear. *Proceedings of the Royal Entomological Society of London (A)* 41: 167–70, plus two plates.

Ruxton, G.D., T.N. Sherratt, and M.P. Speed. 2004. *Avoiding Attack.* Oxford: Oxford University Press.

Schmidt, J.O. 1990. Predation prevention: Chemical and behavioral counterattack. In *Insect Defenses,* ed. D.L. Evans and J.O. Schmidt. Albany: State University of New York Press.

———. 1990. Hymenopteran venoms: Striving toward the ultimate defense against vertebrates. In *Insect Defenses,* ed. D.L. Evans and J.O. Schmidt. Albany: State University of New York Press.

Tehon, L.R., C.C. Morrill, and R. Graham. 1946. *Illinois Plants Poisonous to Livestock.* College of Agriculture Extension Service in Agriculture and Home Economics Circular 599. Urbana: University of Illinois.

Waldbauer, G.P. 1984. A warningly colored fly, *Stratiomys badius* Walker (Diptera: Stratiomyidae), uses its scutellar spines in defense. *Proceedings of the Entomological Society of Washington* 86: 722–23.

Whitman, D.W., M.S. Blum, and D.W. Alsop. 1990. Allomones: Chemicals for defense. In *Insect Defenses,* ed. D.L. Evans and J.O. Schmidt. Albany: State University of New York Press.

Wigglesworth, V.B. 1972. *The Principles of Insect Physiology.* 7th ed. London: Chapman and Hall.

Wilson, E.O. 1994. *Naturalist.* New York: Warner Books.

Yosef, R., and D. Whitman. 1993. An imperfect defense. *Living Bird* (Autumn): 27–29.

## 9. THE PREDATORS' COUNTERMEASURES

Alcock, J. 1993. *Animal Behavior.* 5th ed. Sunderland, MA: Sinauer Associates.

Bent, A.C. 1942. *Life Histories of North American Flycatchers, Larks, Swallows, and Their Allies.* U.S. National Museum, Bulletin 179.

Brandt, H. 1951. *Arizona and Its Bird Life.* Cleveland: privately printed by the Bird Research Foundation.

Brower, L.P., and L.S. Fink. 1985. A natural toxic defense system: Cardenolides in butterflies versus birds. *Annals of the New York Academy of Sciences* 443: 171–88.

Calvert, W.H., L.E. Hedrick, and L.P. Brower. 1979. Mortality of the monarch butterfly (*Danaus plexippus* L.): Avian predation at five overwintering sites in Mexico. *Science* 204: 847–51.

Corcoran, A.J., J.R. Barber, and W.E. Conner. 2009. Tiger moth jams bat sonar. *Science* 325: 325–27.

Dunning, D.C. 1968. Warning sounds of moths. *Zeitschrift für Tierpsychologie* 25: 129–38.

Dunning, D.C., and K.D. Roeder. 1965. Moth sounds and the insect-catching behavior of bats. *Science* 147: 173–74.

Edmunds, M. 1974. *Defence in Animals*. New York: Longman Group.

Fink, L.S., and L.P. Brower. 1981. Birds can overcome the cardenolide defence of monarch butterflies in Mexico. *Nature* 291: 67–70.

Fullard, J.H., M. Brock Fenton, and J.A. Simmons. 1979. Jamming bat echolocation: The clicks of arctiid moths. *Canadian Journal of Zoology* 57: 647–49.

Glendinning, J.I., A.A. Mejia, and L.P. Brower. 1988. Behavioral and ecological interactions of foraging mice (*Peromyscus melanotus*) with overwintering monarch butterflies (*Danaus plexippus*) in Mexico. *Oecologia* 75: 222–27.

Griffin, D.R. 1958. *Listening in the Dark*. New Haven, CT: Yale University Press.

Heatwole, H. 1965. Some aspects of the association of cattle egrets with cattle. *Animal Behaviour* 13: 79–83.

Heinrich, B. 1971. The effect of leaf geometry on the feeding behavior of the caterpillar of *Manduca sexta* (Sphingidae). *Animal Behaviour* 19: 119–24.

———. 1993. How avian predators constrain caterpillar foraging. In *Caterpillars*, ed. N.E. Stamp and T.M. Casey. New York: Chapman and Hall.

Heinrich, B., and S.I. Collins. 1983. Caterpillar leaf damage and the game of hide-and-seek with birds. *Ecology* 64: 592–602.

Hristov, N.I., and W.E. Conner. 2005. Sound strategy: Acoustic aposematism in the bat–tiger moth arms race. *Naturwissenschaften* 92: 164–69.

Marshall, S.A. 2006. *Insects: Their Natural History and Diversity*. Buffalo, NY: Firefly Books.

May, M. 1991. Aerial defense tactics of flying insects. *American Scientist* 79: 316–28.

Peterson, R.T. 1963. *The Birds*. New York: Time.

Roeder, K.D. 1962. The behaviour of free flying moths in the presence of artificial ultrasonic pulses. *Animal Behaviour* 10: 300–304, plus 3 plates.

———. 1963. *Nerve Cells and Insect Behavior.* Cambridge, MA: Harvard University Press.

Roeder, K.D., and A.E Treat. 1961. The detection and evasion of bats by moths. *American Scientist* 49: 135–48.

Shober, W. 1984. *The Lives of Bats.* Translated from the German by S. Furness. London: Croom Helm.

Smith, S.M. 1991. *The Black-Capped Chickadee.* Ithaca, NY: Cornell University Press.

Spangler, H.G. 1988. Hearing in tiger beetles. *Physiological Entomology* 13: 447–52.

Sternburg, J.G., G.P. Waldbauer, and A.G. Scarbrough. 1981. The distribution of the Cecropia moth (Saturniidae) in central Illinois: A study in urban ecology. *Journal of the Lepidopterists' Society* 35: 304–20.

Sullivan, K.A. 1984 Information exploitation by downy woodpeckers in mixed-species flocks. *Behaviour* 91: 294–311.

Tinbergen, L. 1960. The natural control of insects in pinewoods. 1. Factors influencing the intensity of predation by songbirds. *Archives Neerlandaises de Zoologie* 13: 265–343.

Tinbergen, N., M. Impekoven, and D. Franck. 1967. An experiment on spacing-out as a defence against predation. *Behaviour* 28: 307–21. Reprinted in N. Tinbergen, *The Animal in Its World.* Vol. 1. Cambridge, MA: Harvard University Press, 1972.

Treat, A.E. 1955. The response to sound in Lepidoptera. *Annals of the Entomological Society of America* 48: 272–84.

Waldbauer, G.P., and J.G. Sternburg. 1982. Cocoons of *Calosamia promethea* (Saturniidae): Adaptive significance of differences in mode of attachment to the host tree *Journal of the Lepidopterists' Society* 36: 192–99.

## 10. PROTECTION BY DECEPTION

Atkins, E.L., Jr. 1943. Mimicry between the drone-fly, *Eristalis tenax* (L.), and the honeybee, *Apis mellifera* L. Its significance in ancient mythology and present-day thought. *Annals of the Entomological Society of America* 41: 387–92.

Bates, H.W. 1862. Contributions to an insect fauna of the Amazon valley: Lepidoptera: Heliconidae. *Transactions of the Linnean Society of London* 23: 495–566, plus 2 plates.

Borror, D.J., D.M. De Long, and C.A. Triplehorn. 1981. *An Introduction to the Study of Insects.* 5th ed. Philadelphia: Saunders College Publishing.

Bouseman, J.K., and J.G. Sternburg. 2001. *Field Guide to Butterflies of Illinois.* Illinois Natural History Survey Manual 9. Champaign: Illinois Natural History Survey.

Brower, J.V.Z. 1958. Experimental studies of mimicry in some North American butterflies. Part I. The monarch, *Danaus plexippus,* and the viceroy, *Limenitis archippus. Evolution* 12: 32–47.

———. 1958. Experimental studies of mimicry in some North American butterflies. Part II. *Battus philenor* and *Papilio troilus, P. polyxenes,* and *P. glaucus. Evolution* 12: 123–36.

Brower, L.P. 1969. Ecological chemistry. *Scientific American* 220: 4–15.

Brower, L.P., and J.V.Z. Brower. 1960. Experimental studies of mimicry. 5. The reactions of toads *(Bufo terrestris)* to bumblebees *(Bombus americanorum)* and their robberfly mimics *(Mallophora bomboides),* with a discussion of aggressive mimicry. *American Naturalist* 94: 343–55.

———. 1962. Experimental studies of mimicry. 6. The reactions of toads *(Bufo terrestris)* to honeybees *(Apis mellifera)* and their dronefly mimics *(Eristalis vinetorum). American Naturalist* 96: 297–308.

———. 1962. Investigations into mimicry. *Natural History* 71: 8–19.

———. 1965. Experimental studies of mimicry. 8. Further investigation of honeybees *(Apis mellifera)* and their dronefly mimics *(Eristalis). American Naturalist* 99: 173–88.

Brower, L.P., J.V.Z. Brower, F.G. Stiles, H.J. Croze, and A.S. Hower. 1964. Mimicry: Differential advantage of color-patterns in the natural environment. *Science* 144: 183–85.

Brower, L.P., L.M. Cook, and H.J. Croze. 1967. Predator responses to artificial Batesian mimics released in a neotropical environment. *Evolution* 21: 11–23.

Cook, L.M., L.P. Brower, and J. Alcock. 1969. An attempt to verify mimetic advantage in a neotropical environment. *Evolution* 23: 339–45.

Cott, H.B. 1957. *Adaptive Coloration in Animals.* London: Methuen.

Evans, D.L., and G.P. Waldbauer. 1982. Behavior of adult and naive birds when presented with a bumblebee and its mimic. *Zeitschrift für Tierpsychologie* 59: 247–59.

Gamberale-Stille, G., and T. Guilford. 2003. Contrast versus colour in aposematic signals. *Animal Behaviour* 65: 1021–26.

Gaul, A. T. 1952. Audio mimicry: An adjunct to color mimicry. *Psyche* 59: 82–83.

Gilbert, L.E. 1983. Coevolution and mimicry. In *Coevolution*, ed. D.J. Futuyma and M. Slatkin. Sunderland, MA: Sinauer Associates

Grafen, A. 1990. Biological signals as handicaps. *Journal of Theoretical Biology* 144: 517–46.

Jeffords, M.R., J.G. Sternburg, and G.P. Waldbauer. 1979. Batesian mimicry: Field demonstration of the survival value of pipevine swallowtail and monarch color patterns. *Evolution* 33: 275–86.

Linsenmaier, W. 1972. *Insects of the World*. Translated from the German by L.E. Chadwick. New York: McGraw-Hill.

Maier, C.T., and G.P. Waldbauer. 1979. Diurnal activity patterns of flower flies (Diptera: Syrphidae) in an Illinois sand area. *Annals of the Entomological Society of America* 72: 237–45.

———. 1979. Dual mate-seeking strategies in male syrphid flies (Diptera: Syrphidae). *Annals of the Entomological Society of America* 72: 54–61.

Metcalf, R.L., and R.A. Metcalf. 1993. *Destructive and Useful Insects*. 5th ed. New York: McGraw-Hill.

Morgan, C.L. 1896. *Habit and Instinct*. London: Edward Arnold.

Moss, A.M. 1920. Sphingidae of Pará, Brazil. *Novitates Zoologicae* 27: 333–424, plus 11 plates.

Rashed, A., M.I. Khan, J.W. Dawson, J.E. Yack, and T.N. Sherratt. 2009. Do hoverflies (Diptera: Syrphidae) sound like the Hymenoptera they morphologically resemble? *Behavioral Ecology* 20: 396–402.

Ritland, D.B., and L.P. Brower. 1991. The viceroy butterfly is not a Batesian mimic. *Nature* 350: 497–98.

Sternburg, J.G., G.P. Waldbauer, and M.R. Jeffords. 1977. Batesian mimicry: Selective advantage of color pattern. *Science* 195: 681–83.

Tinbergen, N. 1951. *The Study of Instinct*. Oxford: Clarendon Press.

———. 1965. *Animal Behavior*. New York: Time-Life Books.

Waldbauer, G.P. 1970. Mimicry of hymenopteran antennae by Syrphidae. *Psyche* 77: 45–49.

———. 1988. Aposematism and Batesian mimicry: Measuring mimetic advantage in natural habitats. *Evolutionary Biology* 41: 227–59.

———. 1988. Asynchrony between Batesian mimics and their models. In *Mimicry and the Evolutionary Process*, ed. L.P. Brower. Chicago: University of Chicago Press.

Waldbauer, G. P., and W. E. LaBerge. 1985. Phenological relationships of wasps, bumblebees, their mimics, and insectivorous birds in northern Michigan. *Ecological Entomology* 10: 99–110.

Waldbauer, G. P., and J. K. Sheldon. 1971. Phenological relationships of some aculeate Hymenoptera, their dipteran mimics, and insectivorous birds. *Evolution* 25: 371–82.

Waldbauer, G. P., and J. G. Sternburg. 1975. Saturniid moths as mimics: An alternative interpretation of attempts to demonstrate mimetic advantage in nature. *Evolution* 29: 650–58.

———. 1983. A pitfall in using painted insects in studies of protective coloration. *Evolution* 37: 1085–86.

Waldbauer, G. P., J. G. Sternburg, and A. W. Ghent. 1988. Lakes Michigan and Huron limit gene flow between the subspecies of the butterfly *Limenitis arthemis. Canadian Journal of Zoology* 66: 1790–95.

Waldbauer, G. P., J. G. Sternburg, and C. T. Maier. 1977. Phenological relationships of wasps, bumblebees, their mimics, and insectivorous birds in an Illinois sand area. *Ecology* 58: 583–91.

Wickler, W. 1968. *Mimicry in Plants and Animals.* Translated from the German by R. D. Martin. New York: McGraw-Hill.

EPILOGUE

Brower, J. V. Z. 1958. Experimental studies of mimicry in some North American butterflies. Part III. *Danaus gilippus berenice* and *Limenitis archippus floridensis. Evolution* 12: 273–85.

Brower, L. P., ed. 1988. *Mimicry and the Evolutionary Process.* Chicago: University of Chicago Press.

Brower, L. P., J. V. Z. Brower, and C. T. Collins. 1963. Experimental studies of mimicry. 7. Relative palatability and Müllerian mimicry among neotropical butterflies of the subfamily Heliconiinae. *Zoologica* 48: 65–84, plus 1 plate.

DeBach, P. 1974. *Biological Control by Natural Enemies.* Cambridge: Cambridge University Press.

Goldschmidt, R. B. 1945. Mimetic polymorphism, a controversial chapter of Darwinism. *Quarterly Review of Biology* 20: 147–64, 205–30.

Gould, S. J., and N. Eldredge. 1977. Punctuated equilibria: The tempo and mode of evolution reconsidered. *Paleobiology* 3: 115–51.

Guilford, T. 1990. The evolution of aposematism. In *Insect Defenses,* ed. D.L. Evans and J.O. Schmidt. Albany: State University of New York Press.

Mayr, E. 1963. *Populations, Species, and Evolution.* Cambridge, MA: Harvard University Press.

Ruxton, G.D., T.N. Sherratt, and M.P. Speed. 2004. *Avoiding Attack.* Oxford: Oxford University Press.

Strickberger, M.W. 1996. *Evolution.* 2nd ed. Boston: Jones and Bartlett.

# INDEX

| | |
|---:|:---|
| Text: | 10.75/15 Janson MT Pro |
| Display: | Janson MT Pro |
| Compositor: | BookComp, Inc. |
| Indexer: | Thérèse Shere |
| Printer and binder: | Maple-Vail Book Manufacturing Group |